Issues in the Implementation of Digital Feedback Compensators

The MIT Press Series in Signal Processing, Optimization, and Control
Alan S. Willsky, editor

Issues in the Implementation of Digital Feedback Compensators

Paul Moroney

The MIT Press
Cambridge, Massachusetts
London, England

This book was set in Times New Roman
by Asco Trade Typesetting Ltd., Hong Kong,
and printed and bound by Halliday Lithograph
in the United States of America.

Library of Congress Cataloging in Publication Data

Moroney, Paul.
  Issues in the implementation of digital feedback compensators.

  (The MIT Press series in signal processing, optimization, and control; 5)
  Based on the author's thesis (Ph.D.—Massachusetts Institute of Technology, 1979) under title: Issues in the digital implementation of control compensators.
  Bibliography: p.
  Includes index.
  1. Discrete-time systems. 2. Digital control systems. I. Title. II. Series.
QA402.M67 1983   001.64   82-10048
ISBN 0-262-13185-4

Contents

List of Figures
and Tables

Figures

Tables

Series Foreword

The fields of signal processing, optimization, and control stand as well-developed disciplines with solid theoretical and methodological foundations. While the development of each of these fields is of great importance, many future problems will require the combined efforts of researchers in all of them. Among these challenges are the analysis, design, and optimization of large and complex systems, the effective utilization of the capabilities provided by recent developments in digital technology for the design of high-performance control and signal processing systems, and the application of systems concepts to a variety of applications, such as transportation systems, seismic signal processing, and data communication networks.

This series serves several purposes. It not only includes books at the leading edge of research in each field but also emphasizes theoretical research, analytical techniques, and applications that merit the attention of workers in all disciplines. In this way the series should help acquaint researchers in each field with other perspectives and techniques and provide cornerstones for the development of new research areas within each discipline and across the boundaries.

Paul Moroney's book *Issues in the Implementation of Digital Feedback Compensators* is a perfect example of the blending of perspectives and methods from two disciplines. In particular, in this book Moroney investigates the implementation of digital control systems, and in doing this he draws heavily on ideas and methods originally developed in the context of digital filter design. However, his work is far more than a simple application of methods in one discipline to problems in another, and for this reason this book offers something to researchers in both fields. For those whose primary interest is in control theory, Moroney's work provides a thorough treatment of and an appreciation for several problems in control system implementation that have never before been dealt with in anywhere near this depth, and certainly not in such a novel manner. For the researcher in digital filter design and digital signal processing, this book provides a useful introduction to the special issues of importance in designing digital systems for the purpose of control and to the significant modifications and extension of existing techniques that these issues require.

Moroney shows that the specification of a compensator in terms of a Kalman filter and a feedback gain is far from the last step in designing a digital control system. In particular, he examines the effect of finite computer wordlength on the performance achieved using different structures for the implementation of digital compensators and demonstrates that there are far superior alternatives to the straightforward coding of the Kalman filter equations. This development involves a significant extension of the tools of analysis used in digital filtering in order both to use performance measures appropriate for the control context and to take into account that a digital control system is embedded in a feedback loop.

In the process of developing a framework for analyzing digital systems in feedback loops, the author provides an excellent exposition of the fact that timing issues play a far more critical role in the control problem than in digital filtering applications due to the presence of what amounts to a clock, namely, the open-loop plant to be controlled. This leads to several important differences between structures for digital filters and digital compensators. For example, the compensator counterpart of a cascaded direct form I structure, which is never used in digital filter design because of its inefficiency, is a potentially useful *and* efficient realization of a digital control system. Furthermore, in his treatment of pipelined architectures, Moroney demonstrates the necessity of accounting for the temporal dynamics of the resulting digital implementation and shows how this can be done by modifying the original compensator design problem.

Paul Moroney's book represents an important contribution to the subject of digital control system design. Its greatest value, however, may not be in its results but rather in its approach. Much remains to be done in this area, and the ideas and the blending of results and concepts that form the core of this book should be of significant use to others.

Alan S. Willsky

Acknowledgments

This work is based primarily on my doctoral dissertation, "Issues in the Digital Implementation of Control Compensators," submitted to the Massachusetts Institute of Technology September 1979. I would sincerely like to acknowledge the efforts of those who contributed to the original thesis effort and to the current revision.

Technically, I am deeply indebted to my thesis advisors Alan Willsky and Paul Houpt for their direction and encouragement. I would also like to thank my readers Gerry Prado, Gunter Stein, and Jim McClellan for their comments and help during the course of my research.

I am also indebted to the Charles Stark Draper Laboratory for the financial support and the resources that they made available to me during my thesis research, and to the MIT Real Time Systems group for the use of their text-processing system in the preparation of the final thesis document. Much of the efforts of Clark Baker, my brother John, and my wife Jean on that document carried directly over to the current revision. I would also like to thank the LINKABIT Corporation for having the type of environment and the excellent facilities that I needed during the revision and rewrite. For revisions and updates, the reader may write to me at LINKABIT, 3033 Science Park Road, San Diego, CA 92121.

Finally, a sincere thanks to my parents and brother for the support they provided during my research at MIT, and especially to my wife, for her encouragement and patience over the past few months.

1

Introduction

Control theorists have developed many elegant techniques for the design of discrete-time compensators; these include optimal regulators, pole-placement concepts, observer theory, optimal filtering [1–3], and the more traditional methods of classical control [4]. Such compensators are typically designed "off-line" on a large-scale floating-point computer system, where speed and accuracy are more or less assured. The resulting compensator can be so complex, however, that it must also be implemented on a large- or moderate-scale machine (typically expensive and/or slow). In the past, this problem has limited the range of techniques that can be used for applications requiring low-cost or high-speed dedicated controllers. As a result, most existing digital controllers have been quite simple, usually of the proportional-integral-derivative (PID) type [5].

The recent advances in digital hardware capabilities, and in particular the development of the microprocessor, have made it possible to build more complex and more effective real-time digital controllers in an economical manner [5, 6]. However, the use of low-cost digital hardware raises a whole set of issues associated with approximating the near-ideal controller parameters generated off-line. Such issues include computational speed, finite memory constraints (finite precision), arithmetic type, and overall expense. Such issues have not been addressed in the idealized mathematical design procedures developed for control systems.

In implementing such a compensator, our aim is to produce a finite-precision digital system that either performs as close to the ideal (designed) compensator as is consistent with the expense and speed requirements of the application or meets a specific level of performance relative to the ideal as inexpensively as possible subject to certain speed (sampling-rate) constraints. In this investigation we shall use the term *implementation* to refer to (1) the selection of a *structure* — the specification and ordering of the computations that take place in the compensator during each sampling period — and also to (2) the selection of a hardware architecture and digital components. We shall also use the term *ideal* or *infinite precision* when referring to the compensator parameters that are generated off-line, even though this is true only in a relative sense. It is also important to note that the mathematical design technique that produces an ideal compensator and the implementation of this compensator are not necessarily independent processes. For example, when designing a discrete-time controller for a continuous-time plant, a sampling rate must first be assumed. However, the implementation of the resulting controller is frequently very important in determining the maximum sampling rate allowed. Thus interaction between the design and implementation phases may be necessary.

Some effort has been directed toward investigating the issues involved in implementing digital feedback compensators, but it has been somewhat limited. Knowles and Edwards [7] and Curry [8] have considered a roundoff noise analysis of certain sampled-data systems. Bertram [9], Slaughter [10], Johnson [11], and Lack and Johnson [12] have developed amplitude bounds on the effects of quantization in sampled-data control systems. Sripad [13] has looked in some depth at the roundoff noise and finite-precision coefficient performance of the discrete-time Kalman filter and linear-quadratic-Gaussian (LQG) controller. Rink and Chong [14] have derived bounds on the effects of quantization errors in floating-point regulators. Farrar [15] has pointed out in a basic way some of the issues involved in implementing continuous-time LQG controllers as discrete-time fixed-point microprocessor-based systems.

In his monograph, Willsky [16] has discussed a great number of parallels between the fields of digital signal processing and control and estimation. Many of the basic issues involved in implementing digital feedback compensators have been examined in the context of digital signal processing, and a great many results exist. These digital filtering results are very important for control applications since a digital control system can simply be viewed as a digital filter (the compensator) embedded in a feedback loop through a controlled plant. However, only in a few special cases do these results apply directly to control. Our approach will be to *use, adapt,* and *extend* these results to the implementation of digital feedback compensators. In some cases we shall use the filtering results directly. However, much of the time the control setting adds new twists to the implementation issues, requiring the adaptation of existing results. This effort, bridging two disciplines, is the most important contribution of this work. In addition, some of the results developed here extend existing methods or introduce new approaches that are also useful for digital filtering applications. This contribution, although limited in scope, may be of value to researchers in digital signal processing.

In order to develop specific results for digital compensator implementation, we have limited this investigation to a class of control problems, namely, steady-state linear-quadratic-Gaussian (LQG) control problems. Controllers based on this framework have been shown to have desirable performance properties in terms of their robustness, multivariate formulation, optimal nature, and so forth. The LQG problem has also received a great deal of attention in the recent literature and is being increasingly applied to real systems. Furthermore, the LQG problem has an explicit scalar objective function, which can be adopted as a performance metric against which the degradation due to finite wordlength effects can be measured. In fact, this was the degradation measure used by Sripad [13]. It is not necessary to choose this performance metric, or even use an LQG framework, but such a choice allows us to develop our results in a concrete setting. Using this LQG control framework in the context of a *single-input single-output* time-invariant system, we can bring out all the issues we wish to raise. Note that in this context, an ideal, or infinite-precision, compensator can be completely described by its scalar transfer function. Thus our specific task will be to implement this transfer function accurately under certain speed, cost, arithmetic, and finite-memory constraints.

In chapter 2 the details of the discrete-time LQG problem under consideration will be presented. Specifically, we shall consider a continuous-time plant which is driven by additive white Gaussian noise and whose measured output is also corrupted by Gaussian noise and then sampled at rate $1/T$. The ideal discrete-time compensator will minimize an equivalent discrete-time performance index, subject to a piecewise-constant control signal $u[t]$. In presenting the equations for this ideal compensator, an important point will be raised. The finite calculation time implicit in the arithmetic operations of the compensator imposes a limit on the sampling rate of the system. Due to this same finite computation time, a realistically designed compensator must have its output at a given sample time depend only on *past* values of the compensator input. However, we may not wish to wait an entire sample period to apply a control computed from past output values. Consequently, we shall discuss a *sample-skew* approach [1] to solving this problem, which involves sampling the compensator input and output at different times.

One of the important issues in discussing digital implementations is the notion of a *structure*. Given the system sample rate, the effects of finite precision on performance are dependent on the structure chosen, and not on the detailed architecture or components selected. If all compensator computations could be performed with infinite precision, then all structures for implementing a given ideal compensator would be equivalent in performance. However, under the real constraint of finite precision, each structure will in general perform at a different level. Chapter 3 will describe several compensator structures. Two important points will be stressed. First, the state-space notation prevalent in control and estimation is not sufficient to represent all possible compensator structures. Second, the concept of a structure for digital *filters* is not quite the same as the concept of a structure for digital *compensators*. This difference requires adaptation of the notation developed by Chan [17] for the representation of digital filter structures to the control case. A major implication of this change is that an nth-order LQG compensator (for an nth-order system) will have $n + 1$ unit delay elements, and not n as in the case of nth-order filters. An important point will also be raised in chapter 3 concerning the use of the ideal compensator equations resulting from the LQG design procedure *as a computational algorithm*. We can simply view this approach as one possible structure, which we shall call the *simple* structure. We will show that although this structure has been frequently used, more or less by default, it is not usually a good choice due to its large number of coefficient multiplications.

Architectural issues will be treated in chapter 4. The ideas of *serialism* and *parallelism*, the degrees to which processes run sequentially or concurrently, will be presented in terms of the trade-off they embody between compensator calculation time, which sets the maximum sample rate, and hardware complexity and expense. These ideas apply directly to digital compensators—no modifications are necessary. However, the same cannot be said for the application of *pipelining* to control systems. Pipelining alters a structure and its transfer function in a very specific way in order to increase its inherent parallelism and thus increase the maximum sampling rate. The use of pipelining in control systems raises an important issue concerning the interaction between the mathematical design of the ideal compensator and the finite-precision implementation of this ideal. The application of pipelining produces additional (unmodeled) series delay in a compensator. If ignored, this delay will appear in a control system as extra negative phase shift, and perhaps cause instability. The only way to account for this delay accurately will be to augment the discretized plant model and redesign the ideal compensator at the new sampling rate. Then if the same pipelining still applies to the new higher-order ideal compensator, improved performance can result.

In chapters 5, 6, and 7 we shall consider the effects of the finite-memory limitations of inexpensive small-scale digital controllers given specific types of arithmetic operations. Any restriction on memory will necessitate finite precision—the use of compensator coefficients (multipliers) of finite wordlength—and the insertion of quantization or overflow non-linearities following the compensator input analog-to-digital (A/D) converter and following certain multiplications (products) and additions. In all cases, fixed-point arithmetic will be assumed, since this tends to be simpler, faster, and less expensive than floating point. Furthermore, the effects of finite precision tend to be less severe under floating-point arithmetic, so analytical techniques are not as important. With these assumptions, methods must be found for selecting minimum coefficient and signal wordlengths that still result in acceptable levels of performance degradation, that is, in sufficiently small increases in the performance index.

Chapter 5 will treat the *uncorrelated* effects of product and A/D quantization on compensator performance. The major effort is devoted to *roundoff* quantization, since the use of roundoff as opposed to *sign-magnitude truncation* results in lower levels of degradation, and also since roundoff effects can be analyzed in a tractable way. The main results of chapter 5 are the adaptations of the scaling and roundoff noise analysis methods of digital filtering to the compensator case. An important implication concerning set-point LQG configurations and the scaling issue will also be presented. Finally, we shall adapt the minimum roundoff noise filter structures developed by Mullis and Roberts [18] to produce minimum roundoff noise compensator structures.

A sixth-order LQG control system will be introduced to test the roundoff analysis method of chapter 5, and a number of different structures will be evaluated on the basis of their roundoff noise performance. We shall show a significant similarity between the results for these structures and the results for filter structures. However, two differences will arise. First, the potential presence of many real poles in a feedback compensator will complicate the pairing issue for parallel and cascade structures. Digital filters typically have at most one real pole, so the issue of pairing real poles is of no interest. Second, although the simple structure will perform relatively well, there will be two structures with many fewer coefficients that perform even better.

Chapter 6 deals with the effect of finite *coefficient* wordlength on performance. This effect is basically deterministic: Given any set of finite wordlength coefficients, we can compute exactly the resulting performance degradation, that is, the increase in the performance index. However, given a degradation level, it will be much harder to find the set of coefficients with the shortest wordlength that meets or exceeds this degradation level. If we make the common assumption that the ideal values of the coefficients will be *rounded* to finite wordlengths (not truncated or chosen in some other way), then the wordlength determination can be accomplished with repeated evaluations of performance, one per wordlength tested. This procedure must also be repeated for each structure considered. Such a technique can be quite costly in a computational sense. Consequently, our emphasis will be on the use of a *statistical* measure of coefficient wordlength developed for digital filter analysis [19–21]. In the digital filter context, this involves the use of first-order sensitivities with respect to the coefficients of the structure. However, for LQG compensators, all the first-order sensitivities of the performance index will be zero due to the initial formulation and optimal nature of the problem. Thus we shall develop two new statistical estimates using *second-order* sensitivities. In fact, if a digital filter is designed to minimize some differentiable scalar function, then second-order sensitivities must be used for any statistical wordlength estimate based on that function. This would constitute an extension to the results for the implementation of digital filters.

We shall test the analytical procedures developed for coefficient wordlength effects with the same sixth-order control system and structures introduced in chapter 5. Again, we shall show the similarity between our results and the filtering results and demonstrate that other structures with far fewer coefficients perform better than the simple structure. The statistical estimates of wordlength will be compared to the exact word-lengths required to meet a specific degradation level. We shall show that the major advantage in using the statistical estimates is not in the computation time they may save over an iterative deterministic method, but in the fact that they are *continuous* and *differentiable* in nature. This fact allows us to apply iterative gradient minimization techniques to compute minimum coefficient wordlength structures, as described in chapter 8. In this procedure, the bulk of the computations for the statistical estimates need be performed only once.

In chapter 7 we shall review the methods used in dealing with the *correlated* effects of the quantization and overflow nonlinearities present in a structure [22]. Any system including nonlinearities can exhibit oscillations, known as *limit cycles*. In digital filtering, there are three basic approaches to combating such effects. First, we can use a structure that can be shown to have no limit cycles, given a specific type of non-linearity. Second, the amplitude of any limit cycles can be upper bounded, allowing us to select a wordlength large enough to make this amplitude negligible. Finally, if a limit cycle occurs, we can inject enough roundoff noise to break up, or *quench*, the oscillation. Our results in this area for digital compensators are quite limited; however, several observations will be made. First, a control system with an open-loop unstable plant or a plant with an integrator pole must of necessity have a low-amplitude limit cycle. Second, the global feedback loop around the compensator can alter the nature of any limit cycles that would occur in the open-loop compen-sator and may even *cause* limit cycles. This point will be demonstrated for a finite impulse response compensator. (A finite impulse response *filter* is not recursive; therefore it can exhibit no limit cycles.) Finally, it is not clear that limit cycles will occur at all in LQG systems, given the system driving noise and measurement noise present. However, jump phenomena and other correlated noise effects may occur.

Chapter 8 will present a general iterative optimization technique for producing minimum roundoff noise and minimum coefficient wordlength structures. This procedure has been adapted from Chan's optimization method for digital filters [17]. Essentially, this technique allows one to

select a structure with a predetermined number of coefficients and itera-
tively vary those coefficients to minimize some scalar criterion. For LQG
compensators, this criterion could be the increase in the performance
index due to roundoff noise, the increase due to finite wordlength coefficients, or some combination of these two. For the minimization of
roundoff effects, the appropriate modification of Chan's procedure will
be similar to the modification developed in chapter 5 for roundoff
analysis. However, the minimization of coefficient wordlength will require
major changes since the statistical wordlength expression must actually
be minimized, and this involves *second-order* sensitivities. The optimization procedure in chapter 8 will also bring out two useful extensions for
the case of digital filtering. First, in minimizing roundoff noise effects,
our procedure will be more general than Chan's by accounting for the
exact number of roundoff error sources and the location of each one in
the structure. This generalization can be easily added to Chan's method.
Second, we shall set forth some general approaches to selecting which
portion of a given structure to optimize, that is, the portion that will
provide the greatest improvement when optimized. (An unconstrained
optimization of the entire structure usually results in too many coefficient
multipliers.) These guidelines also will apply to digital filter structural
optimization.

Finally, in chapter 9 we shall review the contributions of this monograph, being careful to point out where our results are adaptations and
applications of digital filtering techniques to the problem of implementing
digital compensators, and where our results also constitute extensions to
the digital filtering techniques. We shall also suggest how this work might
be extended to other control problems and other compensator design
techniques. Finally, we shall discuss several areas of future research.

2

The LQG Problem

A specific problem formulation is necessary in order to give a unified treatment of the issues involved in implementing digital compensators. Historically, control theory has developed two different approaches—classical control (primarily a frequency-domain approach) and modern control (primarily a time-domain approach). We have selected the linear-quadratic-Gaussian (LQG) modern control problem for several reasons. The design of LQG systems has received a great deal of attention in recent times [3, 23] due to its advantages for control (a multivariate nature, certain robustness properties [24], and so forth). As will be seen, the analysis of LQG compensators brings out all of the issues that we wish to discuss. Furthermore, the LQG problem has a very natural scalar objective criterion for determining its performance—the cost function J (to be defined), which reflects the weighted state and control fluctuations. Such an objective function makes it quite simple to measure the degradation in performance resulting from any given compensator implementation. The most common criticism of the LQG approach, namely, the difficulty in selecting the parameters of J in some meaningful manner, is much less of a problem in light of the recent developments by Harvey and Stein [25] and Stein [26] that relate frequency-domain design parameters to the selection of the scalar function J. It is hoped that our effort will help make the modern control approach more useful for small-scale low-cost digital systems. However, in principle the issues, approaches, and results developed here apply to any control and/or estimation implementation. This chapter will thus present the set of assumptions inherent in the LQG control problem and describe its discrete-time solution.

Consider a continuous-time plant whose performance is to be improved through feedback. Assume that the nth-order state-space equations (2.1) and (2.2) accurately model the input-output behavior of the plant, including any sensor and actuator dynamics (Brackets will indicate continuous-time quantities, while parentheses will indicate discrete-time quantities.):

$$\dot{x}[t] = Ax[t] + Bu[t] + w_1[t], \tag{2.1}$$

$$y[t] = Cx[t] + w_2[t], \tag{2.2}$$

where the time-invariant system matrix A is $n \times n$, the input gain matrix B is $n \times m$, and the output gain vector C is $p \times n$. The n-vector $x[t]$, m-vector $u[t]$, and p-vector $y[t]$ represent the system states, inputs, and outputs, respectively. The n-vector $w_1(t)$ and p-vector $w_2(t)$ represent uncorrelated white Gaussian noise sources of covariances Ξ_1 and Ξ_2, where $\Xi_2 > 0$. It is further assumed that the performance of the system

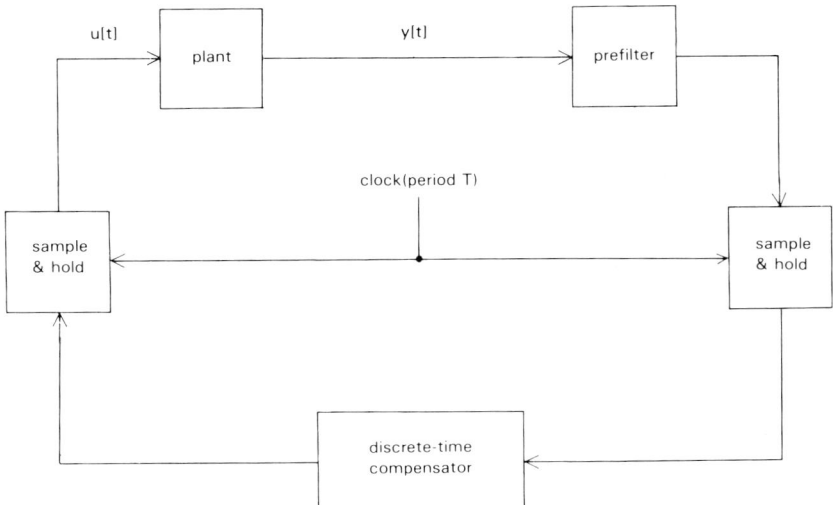

Figure **2.1** LQG configuration.

can be expressed as a scalar quantity that is a quadratic function of the states and controls:

$$J_c = E\left\{\lim_{\tau \to \infty}\frac{1}{2\tau}\int_{-\tau}^{\tau}(x'[t]\hat{Q}x[t] + u'[t]\hat{R}u[t])\,dt\right\}, \tag{2.3}$$

where E represents the expected value operation, the ′ symbol represents the transpose operation, and the weighting matrices $\hat{R}$ and $\hat{Q}$ satisfy $\hat{R} > 0$ and $\hat{Q} \geqslant 0$. We shall deal only with the steady-state LQG problem [1]—hence the time-averaging nature of the performance index in equation (2.3). This formulation also guarantees a time-invariant optimal compensator. The control objective will be to minimize the index J_c with a discrete-time linear compensator as shown in the configuration of figure 2.1, where the input $u[t]$ is now piecewise constant.

The solution to this problem involves discretizing the plant model and performance index and then solving the resulting discrete-time LQG problem. Discretizing the equations (2.1)–(2.3) for a sampling period of T seconds produces [1, 27, 28]

$$x(k + 1) = \Phi x(k) + \Gamma u(k) + w_1(k), \tag{2.4}$$

$$y(k) = Lx(k) + w_2(k), \tag{2.5}$$

$$J_d = E\left\{\lim_{i \to \infty}\frac{1}{2i}\sum_{k=-i}^{i}(x'(k)Qx(k) + 2x'(k)Mu(k) + u'(k)Ru(k))\right\}. \tag{2.6}$$

Note that unlike the continuous time index J_c, the discrete-time index in (2.6) has a cross-term involving the weighting matrix M. Equations (2.4) and (2.5) describe the behavior of the plant at the sample times, and the index J_d in (2.6) satisfies

$$\lim_{T \to 0} J_d = J_c. \tag{2.7}$$

The discrete-time parameters in (2.4)–(2.6) are defined as follows:

$$\Phi(\tau) = e^{A\tau}; \qquad \Gamma(t) = \int_0^t \Phi(\tau)B\,d\tau,$$

$$\Phi = \Phi(T),$$

$$\Gamma = \Gamma(T),$$

$$L = C, \tag{2.8}$$

$$Q = \frac{1}{T} \int_0^T \Phi'(\tau)\hat{Q}\Phi(\tau)\,d\tau,$$

$$R = \hat{R} + \frac{1}{T} \int_0^T \Gamma'(\tau)\hat{Q}\Gamma(\tau)\,d\tau,$$

$$M = \frac{1}{T} \int_0^T \Phi'(\tau)\hat{Q}\Gamma(\tau)\,d\tau.$$

The discrete uncorrelated white noise vectors $w_1(k)$ and $w_2(k)$ have the following covariance matrices:

$$\Theta_1 = \int_0^T \Phi(\tau)\Xi_1\Phi'(\tau)\,d\tau,$$

$$\Theta_2 = \frac{1}{T}\Xi_2. \tag{2.9}$$

The factor of $1/T$ in the expression for Θ_2 arises from the filter preceding the output sampler in figure 2.1. Such a lowpass filter (of bandwidth $2/T$) will be assumed to pass the signal Lx unchanged, while filtering the white measurement noise w_2. Due to the fictitious nature of white noise (its unlimited bandwidth), one cannot actually sample it unfiltered without obtaining a sample of infinite variance. (Aliasing [29–31], which is an overlapping of the spectrum of the sampled signal, would cause the infinite variance.)

The solution to the discrete-time LQG problem, given in Sage [32], gives rise to the following ideal compensator:

$$\hat{x}(k+1) = \Phi\hat{x}(k) + K(y(k+1) - L\Phi\hat{x}(k)) + \Gamma u(k),$$

$$u(k+1) = -G\hat{x}(k+1), \tag{2.10}$$

where $\hat{x}$ represents the state estimate, G is computed off-line as the solution to an optimal regulator problem, and K is computed off-line as the solution to a Kalman filter problem.

Immediately, a problem arises in trying to implement the compensator described in (2.10). The system shown in figure 2.1 and equations (2.4)–(2.6) assumes that the output and input samplers operate simultaneously. However, equations (2.10) clearly show a dependence of $u(k+1)$ on $y(k+1)$. Since it takes a finite amount of time t_c to compute $u(k+1)$ *after* $y(k+1)$ is present at the sampler output, $u(k+1)$ cannot be generated until some time *after* the $(k + 1)$th sample time. This contradiction makes it impossible to implement (2.10) exactly as described.

One way to resolve this problem involves using the configuration and parameters of (2.10) and delaying the clock driving the zeroth-order hold at the compensator output by t_c seconds. Whenever $t_c \ll T$, this approach will result in nearly the same level of performance as would be computed for the nonimplementable (2.10). However, a more general procedure that will work for any $T \geqslant t_c$ is desirable.

Kwakernaak and Sivan [1] have presented such a design method, *including* the possibility of calculation delay in the initial design. Let t_k represent the kth discrete time that the control sample-and-hold unit is sampled and also the kth time at which the states $x[t]$ are discretized. Let t_k'' represent the kth discrete time that the output sample-and-hold unit is sampled. Let us also assume that t_k'' precedes t_{k+1} by the calculation time. Thus the control $u[t_{k+1}]$ will be applied to the plant as soon as it can be computed from the most recent output $y[t_k'']$. In terms of the discrete index k, which is no longer the same time instant for u and y, the control $u(k + 1)$ depends only on observations up to and including $y(k)$. If the calculation time equals t_c, this scheme corresponds to a delaying (skewing) of the clock driving the output sample-and-hold unit by a time $\delta = T - t_c$ relative to the clock driving the system input (u) sample-and-hold (see figure 2.2) [1].

This sampling skew must be reflected in the discrete-time plant model equations [1, section 6.2]. If we rewrite t_k'' as $kT + \delta$, and use the variation-of-constants formula, the output y at t_k'' can be found:

$$y[kT + \delta] = Ce^{A\delta}x[kT] + w_2[kT + \delta]$$

$$+ C\int_{kT}^{kT+\delta} e^{A(kT+\delta-\tau)}(Bu[\tau] + w_1[\tau])\,d\tau \tag{2.11}$$

$$= Ce^{A\delta}x[kT] + w_2[kT + \delta]$$

$$+ C\left\{\int_0^\delta e^{A(\delta-\tau)}\,d\tau\right\}Bu[kT] + C\int_0^\delta e^{A(\delta-\tau)}w_1[\tau]\,d\tau.$$

In its discrete-time form

$$y(k) = Lx(k) + Du(k) + w_2(k), \tag{2.12}$$

where

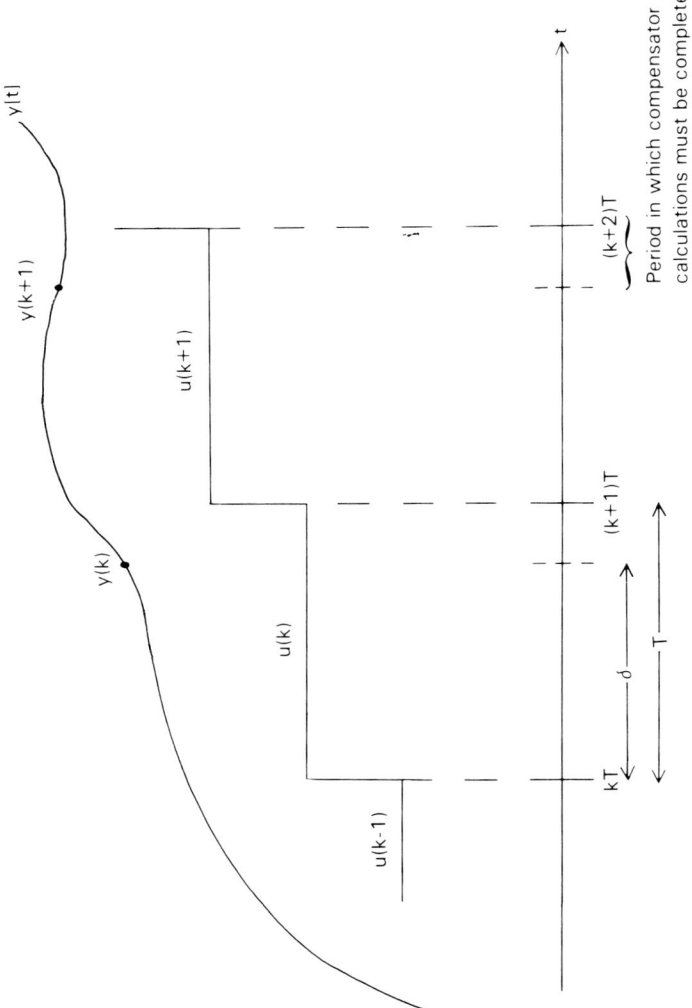

Figure 2.2 Sampling skew (from [1, p. 523]).

$$L = C\Phi(\delta),$$

$$D = C \left[\int_0^\delta \Phi(\delta - \tau)\, d\tau \right] B,$$

$$w_2(k) = w_2[kT + \delta] + C \int_0^\delta \Phi(\delta - \tau) w_1[\tau]\, d\tau.$$

Model equation (2.12) must replace (2.5). Two complications have been introduced: the feedthrough term $Du(k)$ and the nature of the noise $w_2(k)$. The noise vectors $w_1(k)$ and $w_2(k)$ have become correlated due to the difference between the input and output clock phases:

$$E \left\{ \begin{bmatrix} w_1(k) \\ w_2(k) \end{bmatrix} [w_1'(l)\, w_2'(l)] \right\} = \begin{bmatrix} \Theta_{11} & \Theta_{12} \\ \Theta_{12}' & \Theta_{22} \end{bmatrix} \delta_{kl}, \tag{2.13}$$

where

$$\delta_{kl} = \begin{cases} 1 & \text{for } k = l \\ 0 & \text{otherwise}, \end{cases}$$

$$\Theta_{11} = \int_0^T \Phi(\tau) \Xi_1 \Phi'(\tau)\, d\tau,$$

$$\Theta_{22} = \frac{1}{T} \Xi_2 + \int_0^\delta \Phi(\tau) \Xi_1 \Phi'(\tau)\, d\tau,$$

$$\Theta_{12} = \int_0^\delta \Phi(T - \tau) \Xi_1 \Phi'(\delta - \tau)\, d\tau.$$

By restricting $\hat{x}(k + 1)$, or equivalently $u(k + 1)$, to depend on the observations up to and including $y(k)$ only, the optimal compensator equations must also be modified [1]:

$$\hat{x}(k + 1) = \Phi \hat{x}(k) + \Gamma u(k) + K(y(k) - L\hat{x}(k) - Du(k)),$$
$$u(k + 1) = -G\hat{x}(k + 1), \tag{2.14}$$

where K is the new $n \times p$ (steady-state) optimal filter gain matrix and G is the $m \times n$ optimal regulator gain matrix. These matrices satisfy discrete algebraic Riccati equations that can be derived from [1] for the discretized plant and compensator described in (2.4), (2.6), (2.8), and (2.12)–(2.14) (Equation (2.15) is also presented in [26].):

$$\overline{P} = (\Phi - \Gamma R^{-1} M')' \overline{P} (\Phi - \Gamma G) + Q - MR^{-1}M', \qquad (2.15)$$

where

$$G = (R + \Gamma'\overline{P}\Gamma)^{-1} \Gamma'\overline{P}(\Phi - \Gamma R^{-1}M') + R^{-1}M',$$

and

$$\Sigma = (\Phi - KL)\Sigma\Phi' + \Theta_{11} - K\Theta'_{12}, \qquad (2.16)$$

where

$$K = (\Phi\Sigma L' + \Theta_{12})(\Theta_{22} + L\Sigma L')^{-1}.$$

With this formulation, the compensator (2.14) can actually be implemented as long as $0 \leqslant \delta \leqslant T - t_c$ since the time between the reception of $y(k)$ and the generation (sampling) of $u(k + 1)$ would be long enough (at least t_c seconds) to complete the computations involved. Whenever the calculation time is comparable to the sample period, or the sample rate is much greater than the system bandwidth, it is advantageous to choose $\delta = 0$. Such a choice simplifies (2.16) since $\Theta_{12} = 0$, allows for a simpler hardware clocking arrangement for the samplers, and can also reduce the on-line computation time t_c since $D = 0$. In that case, there would only be one system clock; thus all the discrete indices k would indicate the same time instant. For the examples treated in this work, δ will be assumed to be zero for simplicity. The results easily extend to the nonzero δ case.

In this study, only single-input single-output plants will be considered $(m = p = 1)$. With this choice, we can naturally build on the existing digital filtering results and still bring out the issues we wish to discuss. Consideration of the multiple-input multiple-output case would raise even more issues, and probably obscure the points we wish to make. Even in digital signal processing, there are very few multiple-input multiple-output results. The extension of our results for control systems to the multiple-input multiple-output case would be valuable and in most cases is not too difficult. Topics such as multiple-input scaling, multiple-output pipelining, and multiloop limit cycles are discussed in the closing chapter of this work.

For the single-input single-output case, the ideal compensator of (2.14) can be completely described by an input-output map, or transfer function. In terms of the parameters in (2.14), this transfer function can be written

$$H(z) = \frac{U(z)}{Y(z)} = -G(z - \Phi + KL + \Gamma G)^{-1}K. \tag{2.17}$$

When expressed as a ratio of polynomials, equation (2.17) will have the form (2.18), where n is the order of the plant and thus also equals the number of poles in the LQG compensator:

$$H(z) = \frac{a_1 z^{-1} + a_2 z^{-2} + \cdots + a_n z^{-n}}{1 + b_1 z^{-1} + b_2 z^{-2} + \cdots + b_n z^{-n}}. \tag{2.18}$$

The parameters a_i and b_i are functions of the parameters of the LQG control problem. The lack of a constant term a_0 in the numerator of (2.18) follows from the dependence of $u(k)$ only on past values of y as explained in chapter 2. Equation (2.18) can now be taken as a general representation of an ideal, or infinite-precision, LQG compensator. We may now proceed to consider its implementation.

3

Compensator Structures

Chapter 2 has described the background and basic derivation of the LQG compensator. The net result was the set of equations (2.14), which for the single-input single-output case was rewritten as the transfer function (2.17), and then as (2.18). Note that in these transfer functions, y represents the compensator input and u the output, which is the reverse of the filtering case typically considered in digital signal processing. Now consider that (2.17) or (2.18) is to be implemented *digitally* (as a digital network, or filter [33]). Figure 3.1 presents a simple block diagram of this system. The transfer function (3.1) must now be implemented in infinite precision with as little degradation in some system performance measure as possible, subject to certain constraints on the speed and cost of the attendant hardware.

In this chapter, we shall discuss the concept of *structures* for implementing digital compensators and examine accurate ways of representing the arithmetic operations implicit in such structures. Adapting the results of digital signal processing, we shall develop an accurate notation for compensator structures. (Interestingly, the state-space form common in control applications will be shown to be inadequate for this purpose.) Several classes of structures will then be presented using this new notation.

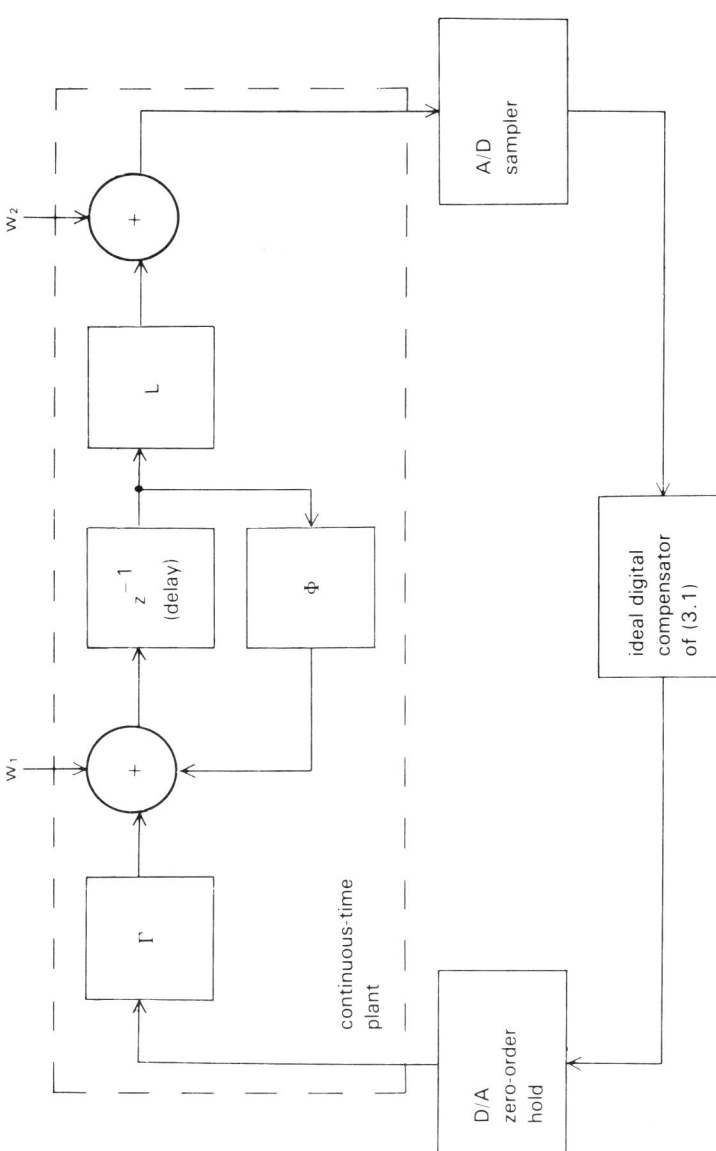

Figure **3.1** Plant and digital compensator.

3.1 Structures

As explained in chapter 1, the term *implementation* includes the choice of a suitable structure to approximate (2.17) [or (2.18)] assuming fixed-point arithmetic and also the specification of the hardware architecture and components. In this section, we shall describe the concept of a structure for implementing a digital filter or compensator.

The term *structure* will be employed to specify the exact finite-precision mathematical procedure by which the compensator (or filter) output samples u are generated from its input samples y. All structures for implementing a given filter or compensator would perform identically under infinite-precision arithmetic, but will in general produce different quantization noise, coefficient quantization effects, and limit cycles given the (realistic) finite-precision environment.

Consider a very simple example. Assume that an ideal compensator has been designed and that its (infinite-precision) transfer function is

$$H(z) = \frac{z^{-1}}{1 + 1.11z^{-1} + 0.287z^{-2}}. \tag{3.1}$$

Figure 3.2a shows a *signal-flow graph* [29,30] of one possible structure, the direct form II [29], for implementing (3.1). The infinite-precision values for b_1 and b_2 can be read directly from (3.1). Given only 10-bit coefficient registers, these values must be quantized (assume rounding). Reserving one bit for the integral portion of the coefficient word (bits to the left of the binary point), one sign bit, and 8 bits for the fractional portion, the rounded coefficient values would be 1.109375 and 0.28515625.

Figure 3.2b shows the flow graph of another common structure, the cascade form. Here we realize (2.18) with a cascade of two first-order sections. The coefficients a_1 and a_2 can be found by factoring the denominator of (2.18). Again, the ideal values must be rounded to fit 10-bit words, producing $a_1 = 0.69921875$ and $a_2 = 0.4105625$.

Now let us examine these two structures given their respective finite-precision coefficients. The (10-bit) direct form II and the cascade have the transfer functions shown in (3.2) and (3.3), respectively:

$$H(z) = \frac{z^{-1}}{1 + 1.109375z^{-1} + 0.28515825z^{-2}}, \tag{3.2}$$

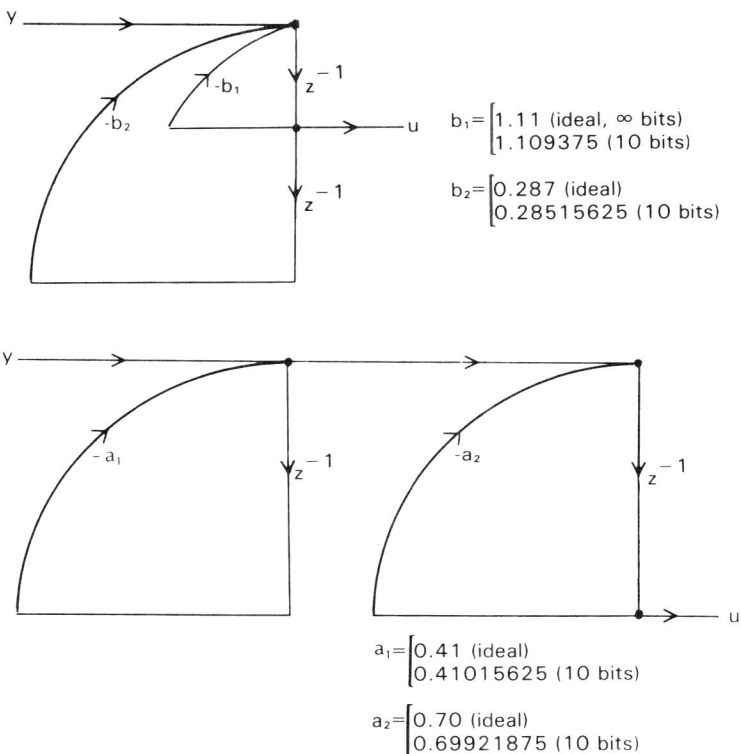

$$b_1 = \begin{bmatrix} 1.11 \text{ (ideal, } \infty \text{ bits)} \\ 1.109375 \text{ (10 bits)} \end{bmatrix}$$

$$b_2 = \begin{bmatrix} 0.287 \text{ (ideal)} \\ 0.28515625 \text{ (10 bits)} \end{bmatrix}$$

$$a_1 = \begin{bmatrix} 0.41 \text{ (ideal)} \\ 0.41015625 \text{ (10 bits)} \end{bmatrix}$$

$$a_2 = \begin{bmatrix} 0.70 \text{ (ideal)} \\ 0.69921875 \text{ (10 bits)} \end{bmatrix}$$

Figure **3.2** Example structures: (a) direct form II; (b) cascade form.

$$H(z) = \frac{z^{-1}}{1 + 1.109375z^{-1} + 0.2867889404296875z^{-2}}. \qquad (3.3)$$

Clearly these two structures produce slightly different transfer functions under finite precision, and thus have slightly different pole locations. They will correspondingly have different quantization noise and limit cycle behavior. Thus different structures will in general have different finite-precision performance, even though their infinite-precision counterparts have equivalent performance (that of the ideal design).

3.2 Structures and Notation

In order to discuss or analyze *different* implementation structures, one must have a notation (other than the pictorial signal-flow graph) that accurately reflects these differences. In this section we shall discuss various methods used to represent digital filter structures, and then we shall develop an adapted notation for accurately describing compensator structures.

From the system-theoretic approach, it seems natural to examine the discrete-time state-space representation for a digital filter (with input y and output u):

$$v(k + 1) = \Psi_{11}v(k) + \Psi_{12}y(k),$$
$$u(k) = \Psi_{21}v(k) + \Psi_{22}y(k). \tag{3.4}$$

In this representation, the states v are defined to be the outputs of the delay elements (state nodes) in a signal-flow graph, and the multiplier coefficients in Ψ_{11}, Ψ_{12}, Ψ_{21}, and Ψ_{22} are the gains between state or input nodes and next-state or output nodes. (In a signal-flow graph, a next-state node is a node that is exited by a delay branch.) Unfortunately, while this form of notation does accurately represent a class of structures, it is not sufficiently general to represent the arithmetic operations associated with any structure. This lack of generality arises in representing structures whose signal-flow graphs must have *intermediate* nodes, that is, nodes that are not state nodes, next-state nodes, or the input or output node. Figure 3.3 presents the signal-flow graph of such a structure, a two-pole two-zero direct form II filter structure. Nodes C and D are

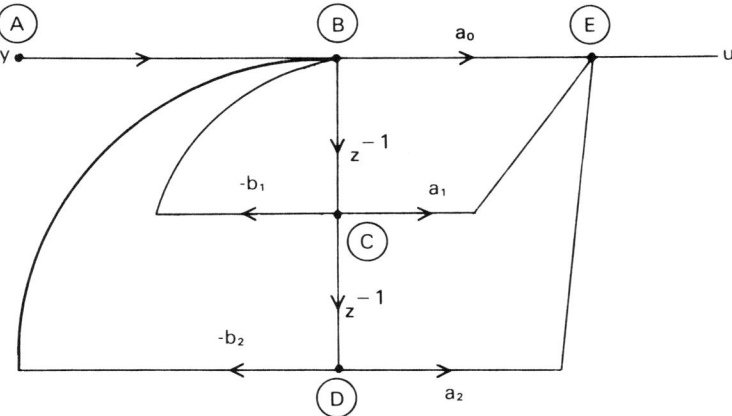

Figure **3.3** Direct form II structure.

state nodes, node A is the input node, and node E is the output node. However, the a_0 branch begins at an intermediate node, node B. Thus there would be no way to include the coefficient a_0 as an entry in any of the state-space matrices Ψ_{11}, Ψ_{12}, Ψ_{21}, or Ψ_{22}. From another viewpoint, the state-space representation lacks any way of expressing the implicit ordering, or *precedence*, associated with the operations involved in certain filter structures. If a structure could be accurately represented with a state-space representation, then the multiplications in that structure could be performed concurrently (and thus independently), and then all the additions could be performed. However, for the direct form II structure of figure 3.3, the multiplications by $-b_1$ and $-b_2$ must precede the addition at node B, which must then precede the following multiplication by a_0. This sequence of operations cannot be adequately expressed by equations of the form (3.4). This fact is clearly illustrated by Willsky [16, pp. 122–124].

At this point it is convenient to turn to the field of digital signal processing for an adequate way to represent structures. Crochiere [34] and Crochiere and Oppenheim [33] have described matrix equations for correctly computing the node signal values in any filter structure. Let the signal value at the ith node (of N_0 nodes) at time k be $u_i(k)$ and the external input to node i be $y_i(k)$. Between any two nodes i and j there can exist one interconnecting *coefficient* branch with gain F_{cij}, and/or one *delay* branch with gain F_{dij}. These branches and their interconnected nodes form an *elementary* network. (We have further assumed that all values F_{dij} are either zero or one, with no loss of generality.) For an elementary network, then, the node value $u_i(k)$ may in general depend on all node values at time $k - 1$ and some of the node values at time k, depending on whatever branches exist:

$$u_i(k) = \sum_{j=1}^{N_0} F_{cji} u_i(k) + \sum_{j=1}^{N_0} F_{dji} u_i(k-1) + y_i(k). \tag{3.5}$$

Thus F_c is an $N_0 \times N_0$ matrix of gains from the coefficient branches, and F_d is an $N_0 \times N_0$ matrix of gains from the delay branches. In most networks, a substantial number of the entries in F_c and F_d are zero, and as stated, the remaining entries in F_d are ones. In z-transform notation, the vector quantity $U(z)$ can be written

$$U(z) = Y(z) + F_c' U(z) + F_d' U(z) z^{-1}. \tag{3.6}$$

The transfer function matrix $H(z)$ defined by $U(z) = H'(z) Y(z)$ can be derived from (3.6):

$$H(z) = [I - F_c - F_d z^{-1}]^{-1}. \tag{3.7}$$

Now let us consider the manner in which the node signal values are computed. These calculations must occur between the time instants $k - 1$ and k. Some of the node updates will involve the past values at time $k - 1$, and some will involve already updated values. Thus the node values must be computed in the proper order. For example, the first node value to be updated should not depend on any other updated node values, since these would not yet have been computed. Thus in terms of the matrix notation, a correct *node precedence*, or ordering, would only depend on the coefficient branches, and not the delay branches, since all the (delay branch) values $u(k)$ can be obtained directly from the known values $u(k - 1)$. Crochiere [34] describes a formal node-ordering technique:

1. All nodes entered by inputs or delay branches *only* are placed in node class 1.

2. Remove from the network all class 1 nodes and any branches entering them.

3. Repeat steps 1 and 2 on the remaining subnetwork, for node classes 2, 3, . . . , until all nodes are classified.

4. Order from 1 to N_0 all nodes, first using all the class 1 nodes, then class 2, and so on.

This technique will not result in a unique ordering of the nodes, but the ordering produced will satisfy the computational constraints mentioned above.

If this ordering procedure can be carried out, the digital network, or structure, is *computable*, and the resulting F_c' matrix is zero on and above the main diagonal. If not, the signal-flow graph has at least one closed path that does not include a delay branch, and thus does not represent an implementable structure. Note that a nonrecursive structure has an ordering whereby F_d' is also zero on and above the main diagonal.

As an example of this matrix signal-flow graph formulation, consider the five-node structure of figure 3.3. Using Crochiere's ordering algorithm, nodes C and D fall into class 1, node A falls into class 2, node B into class 3, and node E into class 4. Thus we can define nodes #1 through

#5 with the ordering C, D, A, B, E. The following five equations now define the (frequency) response of the network:

$$
\begin{aligned}
U_1 &= & & z^{-1} U_4 \\
U_2 &= z^{-1} U_1 \\
U_3 &= -b_1 U_1 - b_2 U_2 & & + Y_3 \\
U_4 &= & U_3 \\
U_5 &= a_1 U_1 + a_2 U_2 & & + a_0 U_4.
\end{aligned}
\tag{3.8}
$$

The 5×5 matrices F_c and F_d can be formed using (3.6) and (3.8), and the resulting matrix $H(z)$ is given in (3.9):

$$
H(z) = \begin{bmatrix}
1 & -z^{-1} b_1 & 0 & -a_1 \\
0 & 1 & b_2 & 0 & -a_2 \\
0 & 0 & 1 & -1 & 0 \\
-z^{-1} & 0 & 0 & 1 & -a_0 \\
0 & 0 & 0 & 0 & 1
\end{bmatrix}^{-1} .
\tag{3.9}
$$

For a single-input single-output digital filter such as the one in figure 3.3, we specify only the scalar input-output map $H_{ij}(z)$ [$H_{35}(z)$ in the example]. The remaining entries of $H(z)$ represent transfer functions from or to nodes that are internal to the structure.

A deficiency of the matrix notation above appears when we consider structural *transformations*. Such transformations are very useful in generating new structures with identical infinite-precision transfer functions as some original structure, but with different finite-precision performance. For a structure that can be accurately represented with state-space notation, the similarity transform fills this role. For the Crochiere matrix representation, a transformation technique also exists [17]. This technique must be constrained so that the transformed structure is computable; in other words, it must have no delay-free loops [17]. However, even with this restriction, the number of delay branches and the degree of precedence inherent in the additions and multiplications of the resulting (infinite-precision-equivalent) structure are in general *unpredictable*.

To combat this difficulty, a notation as convenient and useful for transformation as the state-space form, but with the generality of the Crochiere matrix representation, is desirable. Such a notation, related to the state-space notation, has been presented by Chan [17]. As in a state space, let us define the outputs of delay (storage) elements to be the states v, and let y be the filter or compensator input and u the output. The coefficients and the sequence of multiplications and additions in *any* filter structure can then be specified with the following representation:

$$\begin{bmatrix} v(k+1) \\ u(k) \end{bmatrix} = \Psi_q \Psi_{q-1} \cdots \Psi_1 \begin{bmatrix} v(k) \\ y(k) \end{bmatrix}, \tag{3.10}$$

where $\Psi_q, \ldots, \Psi_1$ are matrices representing the arithmetic and quantization operations in the structure. Three important points make (3.10) useful:

1. Each (finite-precision) coefficient in the structure occurs once and only once as an entry in one of the Ψ_i matrices. The remainder of the matrix entries are ones and zeros.

2. All intermediate (nonstorage) nodes in a structure are represented in the vectors

$$r_1(k) = \Psi_1 \begin{bmatrix} v(k) \\ y(k) \end{bmatrix},$$

$$r_2(k) = \Psi_2 r_1(k), \ldots, \qquad r_{q-1}(k) = \Psi_{q-1} r_{q-2}(k). \tag{3.11}$$

This point is especially important since both the state (storage) nodes v and intermediate nodes r must be scaled to satisfy dynamic range constraints (see chapter 5).

3. The concept of *precedence* for the operations (multiplies, adds, and quantizations) is maintained. The ordering of the Ψ_i matrices implies that the operations involved in computing $r_1(k)$ are completed first, then $r_2(k)$ next, and so forth. Thus the matrix Ψ_q contains the operations of lowest precedence, and the parameter q specifies the number of *precedence levels*.

Chan [17] has also presented a three-step procedure for converting from a signal-flow graph to the representation of (3.10). Step one requires that we add to the signal-flow graph just enough extra nodes and unity gain coefficient branches so that

1. the input node has no branch entering it,

2. the output node has no branch exiting it,

3. any node entered by a delay branch has no other branch entering it,

4. any node exited by a delay branch has no other branch exiting it, and

5. every directed path from an input or state node to an output or next-state node that does not include a delay branch includes the same minimal number q of branches (precedence levels).

Step 2 involves a partitioning of the nodes in the resulting signal flow graph into classes r_k, for $1 \leqslant k \leqslant q + 1$, as follows:

1. The set r_1 consists of all state nodes and input nodes; and
2. The set r_k, for $2 \leqslant k \leqslant q + 1$, consists of all nodes reached from some node in r_{k-1} by a coefficient branch.

Finally, in step 3, we must number the nodes in sets r_1 and r_{q+1} in such a way that within each set the input or output nodes are numbered last. Within all other sets the nodes may be numbered arbitrarily. The matrix Ψ_k can then be defined as the matrix whose (i,j)th element is the gain of the coefficient branch connecting node j in set r_k to node i in set r_{k+1}.

This technique will not produce a unique representation for a signal-flow graph due to the different ways of satisfying step 1 and due to the arbitrary node numbering possible in step 3. (The Crochiere matrix notation is also nonunique due to its arbitrary node numbering within classes.) However, each such representation can be converted back to a signal-flow graph that is equivalent and thus reducible to the original [17]. Consequently, in terms of finite wordlength performance, the different possibilities represent the same structure.

Consider the example of figure 3.2. Using the procedure described above, the direct form II structure in figure 3.2 has a one-level representation as shown in (3.12), while the cascade structure of figure 3.2 requires two levels to describe its operations [see equation (3.13)]:

$$\begin{bmatrix} v(k+1) \\ u(k) \end{bmatrix} = \begin{bmatrix} 0 & 1 & 0 \\ -b_2 & -b_1 & 1 \\ 0 & 1 & 0 \end{bmatrix} \begin{bmatrix} v(k) \\ y(k) \end{bmatrix}, \tag{3.12}$$

$$\begin{bmatrix} v(k+1) \\ u(k) \end{bmatrix} = \begin{bmatrix} 1 & 0 \\ 1 & -a_2 \\ 0 & 1 \end{bmatrix} \begin{bmatrix} -a_1 & 0 & 1 \\ 0 & 1 & 0 \end{bmatrix} \begin{bmatrix} v(k) \\ y(k) \end{bmatrix}. \tag{3.13}$$

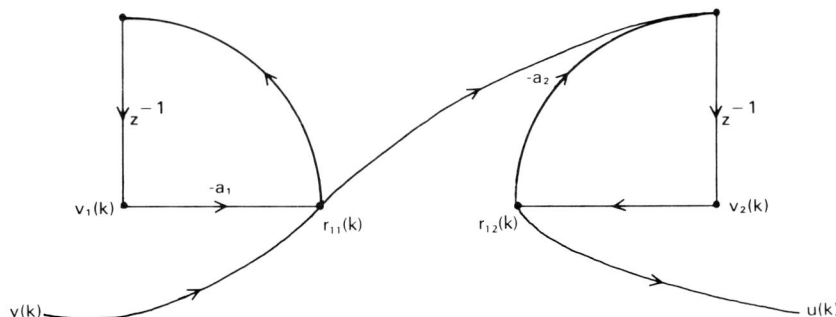

Figure **3.4** Signal-flow graph from (3.13).

Note that if we reverse the procedure and generate a signal-flow graph directly from (3.13), we do not produce the original signal-flow graph (compare figures 3.2b and 3.4). However, nodes $r_{11}(k)$ and $v_1(k+1)$ in figure 3.4 are equivalent nodes, separated only by a trivial multiplication by one in Ψ_2. The same is true of $r_{12}(k)$ and $v_2(k)$. Thus figure 3.2b is simply a *node-minimal* version of figure 3.4 [17, 34]. To summarize, a signal-flow graph for a structure can be converted to several Chan representations, each of which can be converted back to another signal-flow graph. However, each of these graphs can be reduced to the same node-minimal form.

In terms of its generality, the notation described by Chan is as useful as the Crochiere representation. In fact, in the context of Chan's notation, we can now see that a state space will represent only a *class* of structures, namely, those with only one inherent level of precedence.

An important advantage to the notation introduced by Chan is the ease with which transformations [17] can be applied to generate new structures that are infinite-precision-equivalent to some original structure. This technique is an adaption of the similarity transformation used with a (one-level) state space. Using the superscript circle to denote the transformed parameters, let us define

$$\mathring{\Psi}_i = P_i \Psi_i P_{i-1}^{-1} \qquad \text{for } i = 1, \ldots, q, \tag{3.14}$$

where the P_i for $i = 1, \ldots, q-1$ are general nonsingular transformation matrices of appropriate dimension and

$$P_0 = \begin{bmatrix} P & 0 \\ 0 & 1 \end{bmatrix}, \qquad P_q = \begin{bmatrix} P & 0 \\ 0 & 1 \end{bmatrix}. \tag{3.15}$$

The new (transformed) structure will then have the following representation:

$$\begin{bmatrix} \mathring{v}(k+1) \\ u(k) \end{bmatrix} = \mathring{\Psi}_q \mathring{\Psi}_{q-1} \cdots \mathring{\Psi}_1 \begin{bmatrix} \mathring{v}(k) \\ y(k) \end{bmatrix}. \tag{3.16}$$

What makes this transformation method so useful is that the original and transformed structures have the *same* number of states (delays) and the *same* number of precedence levels. It is also possible to restrict the matrices $P_0, P_1, \ldots, P_q$ to control the number of nonunity, nonzero entries in the new transformed modified state-space matrices, as explained in chapter 8.

Now let us try to apply the signal-flow graph and Chan's notation to represent structures for digital feedback compensators. Consider the direct form II structure in figure 3.3 to be a compensator structure by dropping the a_0 branch as implied in (2.18). As required in the compensator formulation of chapter 2, the output at time k does not depend on the input at time k. However, the resulting structure still does not accurately represent the computations that must go on in the compensator. Specifically, the output $u(k)$ seems to depend directly on the values of the state nodes C and D at time k. Yet some computation time must exist to perform the multiplications by a_1 and a_2. Again, for digital filter representations the extra series delay required to complete these multiplications can be ignored. However, for compensator representations, we must be accurate. Figure 3.5 shows how the signal-flow graph in figure 3.3 can be (equivalently) redrawn if the a_0 branch is not present. Now we can see that the value of the output at time k depends only on computations based on state node values at time $k - 1$. Both the graphs of figures 3.3 and 3.5 still represent exactly the same transfer function; however, figure 3.5 accurately shows how the computations in the structure must be organized when we implement this transfer function.

This example points out a difference between signal-flow graphs that represent filter structures and those that represent compensator structures. To represent a compensator structure (that is, its computational algorithm) accurately, the output node must be the output of a delay branch. In other words, the output node must itself be a state node. This requirement can always be met by structures that implement the transfer function

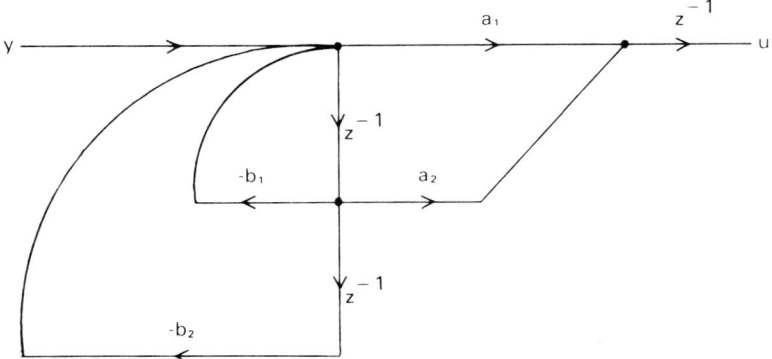

Figure **3.5** Direct form II structure without a_0 branch.

of (2.18). Another implication of this altered notion of a compensator structure is that nth-order compensators will now require structures having at least $n + 1$ unit delay elements (hardware storage registers), rather than n, as with digital filters of order n.

The notation described by Chan is quite general—it is certainly adequate for representing either the signal-flow graph of figure 3.3, or 3.5. Unfortunately, this very generality can lead to problems when we apply structural transformations. Since the Chan representation is so general, transformed structures will tend to contain branches that imply a dependence of the output at time k on state nodes at time k. Rather than require an additional set of constraints on the transformation procedures and notation developed by Chan, let us make the notation itself explicit and thus more clear. Since the output node will be a state, let us define it to be the last state (numerically) and rewrite the Chan notation in the following form:

$$\begin{bmatrix} v(k+1) \\ u(k+1) \end{bmatrix} = \Psi_q \Psi_{q-1} \cdots \Psi_1 \begin{bmatrix} v(k) \\ u(k) \\ y(k) \end{bmatrix}, \qquad (3.17)$$

where the vector v and also the scalar u are the states of the structure. Now, no matter how we transform the Ψ matrices, the output u will explicitly remain a state; in other words, (3.17) is a completely general representation for *compensator* structures. The notation in (3.17) will be called the *modified state-space representation* for a compensator. In order to generate a modified state-space representation from a compensator signal-flow graph, we can still basically follow the three-step procedure outlined by Chan. However, in step 1, we must replace the requirement that no branch exit from the output node with the requirement that the output node be a state node. As described in (3.11), we shall continue to refer to the intermediate nodes as r_i, where $r_1 = \Psi_1[v(k)\ u(k)\ y(k)]'$, $r_2 = \Psi_2 r_1$, and so on.

One final implication of the adapted concept of a structure should be brought out. In terms of the transformation procedure described in (3.14) and (3.15), a change is necessary to accommodate compensator structures. In (3.15), due to the inclusion of the output as a state, the transformation matrix P_0 must now be written

$$\begin{bmatrix} P & 0 & 0 \\ 0 & 1 & 0 \\ 0 & 0 & 1 \end{bmatrix}. \tag{3.18}$$

The extra row and column in this matrix reflect the modified state-space representation, and the unity diagonal entry is necessary since the transformation procedure cannot be permitted to alter the output node.

It is also notationally convenient to define the matrix Ψ_∞. Let the coefficients in each Ψ_i matrix be replaced by their infinite-precision counterparts (their values *before* rounding). Then Ψ_∞ is defined to be the infinite-precision product $\Psi_q \Psi_{q-1} \cdots \Psi_1$. This matrix will be used in the derivations of chapters 5, 6, and 8.

3.3 Classes of Structures

Before discussing some of the various classes of structures that exist, it is important to understand the different points of comparison that should be considered. Beyond the finite wordlength effects of quantization noise, coefficient rounding, and limit cycles that are treated in chapters 5–7, one must compare the number of delay elements, coefficients (multiplications), additions, and precedence levels, and also the number of scalers needed to satisfy dynamic range constraints. We shall examine structures that are typically *canonic* (minimal) with respect to the number of delay elements; such structures thus have a minimal number of storage registers. Also, in order to present specific examples of structures, let us assume that the plant is sixth order ($n = 6$).

Certain common digital filter structures (for example, the direct form II and cascade and parallel structures based on it) will no longer appear quite the same when used for digital compensators. At the very least, each will have an extra delay at the output node. This section will show examples of such structures, and we shall still refer to them by their corresponding digital filtering designations (see figures 3.6, 3.7, and 3.9). For the remainder of this work, the modified state space of (3.17) will be employed to describe compensator structures and all signal-flow graphs will reflect the delay (state) necessary for $u(k)$.

Given the transfer function (2.18), the most straightforward structure to consider is the direct form II [29]. As an LQG compensator structure, its signal-flow graph is shown in figure 3.6. It is canonic in delays with 7 (in general, $n + 1$), has 12 coefficients (nonunity multipliers), and requires only one additional *scaler*. (Scaling, fully discussed in chapter 5, involves a *normalization* of the structure so that roundoff noise effects and overflows can be held to a minimum. In this process, some of a structure's coefficients will be altered, including certain *unity* entries. Such a unity entry will be called a scaling multiplier, or scaler, and indicated in signal-flow graphs and equations with an asterisk.) The coefficients in this structure (before scaling) are read directly from the transfer function (2.18). The modified state space for the direct form II is given below with its two precedence levels:

$$
\Psi_2 =
\begin{bmatrix}
1 & 0 & 0 & 0 & 0 & 0 \\
0 & 1 & 0 & 0 & 0 & 0 \\
0 & 0 & 1 & 0 & 0 & 0 \\
0 & 0 & 0 & 1 & 0 & 0 \\
0 & 0 & 0 & 0 & 1 & 0 \\
0 & 0 & 0 & 0 & 0 & 1 \\
a_6 & a_5 & a_4 & a_3 & a_2 & a_1
\end{bmatrix},
$$

$$
\Psi_1 =
\begin{bmatrix}
0 & 1 & 0 & 0 & 0 & 0 & 0 & 0 \\
0 & 0 & 1 & 0 & 0 & 0 & 0 & 0 \\
0 & 0 & 0 & 1 & 0 & 0 & 0 & 0 \\
0 & 0 & 0 & 0 & 1 & 0 & 0 & 0 \\
0 & 0 & 0 & 0 & 0 & 1 & 0 & 0 \\
-b_6 & -b_5 & -b_4 & -b_3 & -b_2 & -b_1 & 0 & 1^*
\end{bmatrix}.
$$

(3.19)

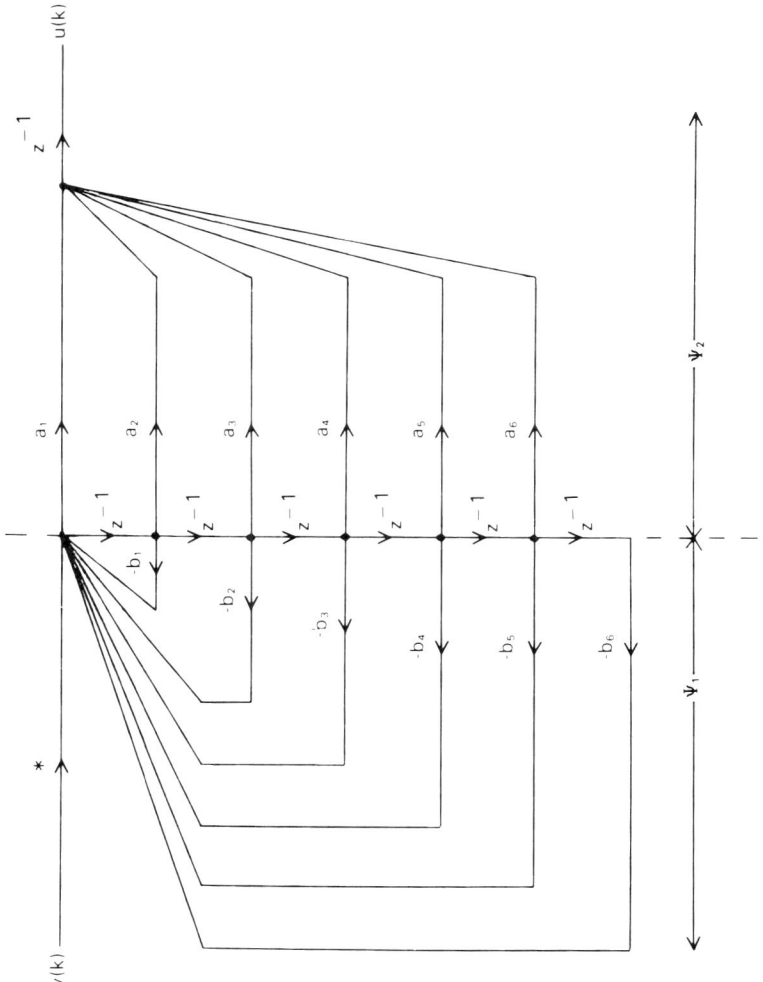

Figure **3.6** Direct form II structure (sixth order).

Figure 3.6 includes a rough indication of which operations belong in which precedence level. Also note that the next-to-last column of Ψ_1 is all zero. This will be true for any structure that has no branches exiting from the output (state) node.

For higher-order filters, the direct form structure is known to perform poorly in terms of the degradation resulting from the use of finite word-lengths [35]. The dynamic range of the coefficients alone grows with filter order, when the poles are clustered in the z plane. (As shown in chapters 5 and 6, this will be true for the direct form II compensator structure also.) Consequently, *factored* structures, such as the *cascade* (of first- and second-order filter sections) are commonly used. This structure is obtained from a multiplicative factoring of the transfer function (2.18):

$$H(z) = \frac{(d_1 z^{-1} + d_2 z^{-2})(1 + d_3 z^{-1} + d_4 z^{-2})(1 + d_5 z^{-1} + d_6 z^{-2})}{(1 + c_1 z^{-1} + c_2 z^{-2})(1 + c_3 z^{-1} + c_4 z^{-2})(1 + c_5 z^{-1} + c_6 z^{-2})}.$$

(3.20)

If each second-order section is implemented as a direct form II structure, then the cascade compensator structure (figure 3.7) also has 12 coefficients and 7 delays (canonic), but requires four precedence levels ($n_s + 1$ in general, where n_s is the number of sections) and three scalers:

$$\Psi_4 = \begin{bmatrix} 0 & 1 & 0 & 0 & 0 & 0 & 0 \\ 1 & 0 & 0 & 0 & 0 & 0 & 0 \\ 0 & 0 & 0 & 1 & 0 & 0 & 0 \\ 0 & 0 & 1 & 0 & 0 & 0 & 0 \\ 0 & 0 & 0 & 0 & 0 & 1 & 0 \\ 0 & 0 & 0 & 0 & 0 & 0 & 1 \\ 0 & 0 & 0 & 0 & d_6 & d_5 & 1^* \end{bmatrix},$$

(3.21)

$$\Psi_3 = \begin{bmatrix} 1 & 0 & 0 & 0 & 0 & 0 & 0 \\ 0 & 1 & 0 & 0 & 0 & 0 & 0 \\ 0 & 0 & 0 & 0 & 0 & 0 & 1 \\ 0 & 0 & 0 & 1 & 0 & 0 & 0 \\ 0 & 0 & 0 & 0 & 1 & 0 & 0 \\ 0 & 0 & 0 & 0 & 0 & 1 & 0 \\ 0 & 0 & d_4 & d_3 & -c_6 & -c_5 & 1^* \end{bmatrix},$$

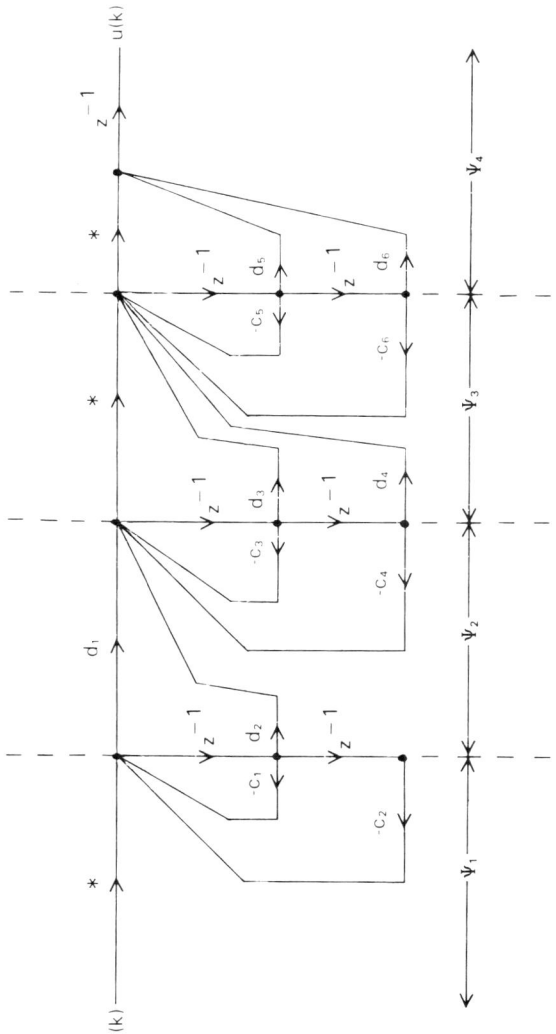

Figure 3.7 Cascade structure (direct form II).

$$\Psi_2 = \begin{bmatrix} 0 & 0 & 0 & 0 & 0 & 1 \\ 1 & 0 & 0 & 0 & 0 & 0 \\ 0 & 1 & 0 & 0 & 0 & 0 \\ 0 & 0 & 1 & 0 & 0 & 0 \\ 0 & 0 & 0 & 1 & 0 & 0 \\ 0 & 0 & 0 & 0 & 1 & 0 \\ d_2 & -c_4 & -c_3 & 0 & 0 & d_1 \end{bmatrix},$$

$$\Psi_{1'} = \begin{bmatrix} 0 & 1 & 0 & 0 & 0 & 0 & 0 \\ 0 & 0 & 1 & 0 & 0 & 0 & 0 \\ 0 & 0 & 0 & 1 & 0 & 0 & 0 \\ 0 & 0 & 0 & 0 & 1 & 0 & 0 \\ 0 & 0 & 0 & 0 & 0 & 1 & 0 \\ -c_2 & -c_1 & 0 & 0 & 0 & 0 & 1^* \end{bmatrix}.$$

Actually, this cascade can be used to represent several different structures since the poles and zeros in (3.20) must first be grouped together to form second-order sections, and then the sections must be ordered. Furthermore, the individual sections could be structured in any number of ways (other than the direct form II) [36,37], each giving rise to a different overall structure.

The consideration of second-order sections other than the direct form II brings out an interesting point. If a cascade or parallel combination of a certain type of section is not delay-canonic when used for digital filters, it may still be delay-canonic when adapted as a compensator structure. Consider the case of a cascade of direct form I [29] second-order sections. Such a *filter* structure is not delay-canonic (it requires more than n delays). However, when adapted to be a *compensator* structure (see figure 3.8), the direct form I structure *is* delay-canonic since it has just $n + 1$ unit delay elements. For a sixth-order LQG compensator, such a structure has 7 delay elements and only 3 (in general n_s) precedence levels and 2 scalers:

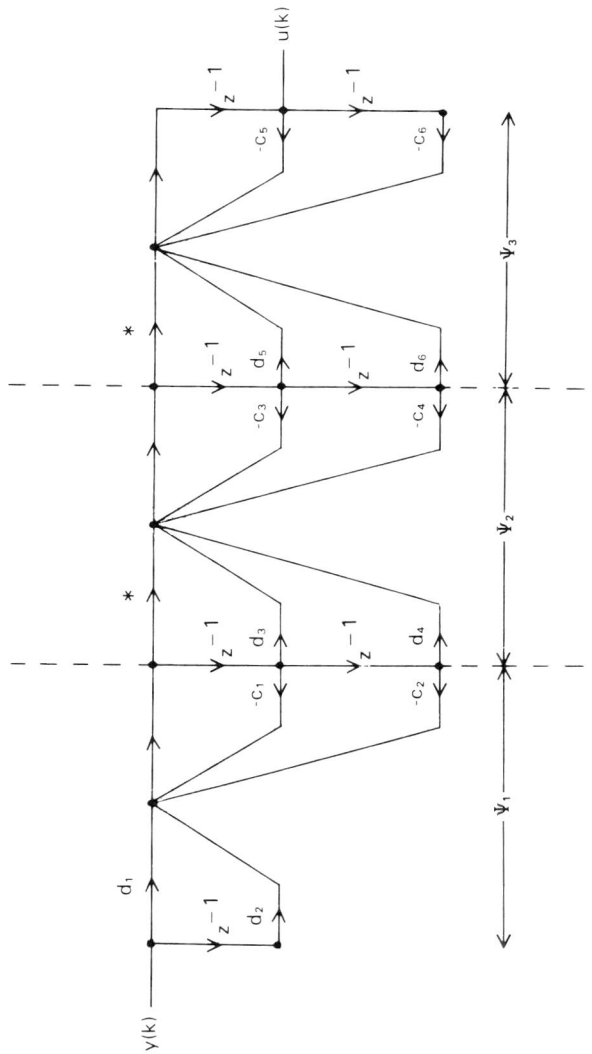

Figure **3.8** Cascade structure (direct form I).

$$\Psi_3 = \begin{bmatrix} 0 & 0 & 0 & 0 & 0 & 1 & 0 & 0 \\ 1 & 0 & 0 & 0 & 0 & 0 & 0 & 0 \\ 0 & 0 & 0 & 0 & 0 & 0 & 1 & 0 \\ 0 & 0 & 1 & 0 & 0 & 0 & 0 & 0 \\ 0 & 0 & 0 & 0 & 0 & 0 & 0 & 1 \\ 0 & 0 & 0 & 0 & 1 & 0 & 0 & 0 \\ 0 & d_6 & d_5 & -c_6 & -c_5 & 0 & 0 & 1^* \end{bmatrix},$$

$$\Psi_2 = \begin{bmatrix} 0 & 1 & 0 & 0 & 0 & 0 & 0 & 0 \\ 0 & 0 & 1 & 0 & 0 & 0 & 0 & 0 \\ 0 & 0 & 0 & 1 & 0 & 0 & 0 & 0 \\ 0 & 0 & 0 & 0 & 1 & 0 & 0 & 0 \\ 0 & 0 & 0 & 0 & 0 & 1 & 0 & 0 \\ 0 & 0 & 0 & 0 & 0 & 0 & 1 & 0 \\ 0 & 0 & 0 & 0 & 0 & 0 & 0 & 1 \\ d_4 & d_3 & -c_4 & -c_3 & 0 & 0 & 0 & 1^* \end{bmatrix}, \qquad (3.22)$$

$$\Psi_1 = \begin{bmatrix} 0 & 1 & 0 & 0 & 0 & 0 & 0 & 0 \\ 0 & 0 & 1 & 0 & 0 & 0 & 0 & 0 \\ 0 & 0 & 0 & 1 & 0 & 0 & 0 & 0 \\ 0 & 0 & 0 & 0 & 1 & 0 & 0 & 0 \\ 0 & 0 & 0 & 0 & 0 & 1 & 0 & 0 \\ 0 & 0 & 0 & 0 & 0 & 0 & 1 & 0 \\ 0 & 0 & 0 & 0 & 0 & 0 & 0 & 1 \\ d_2 & -c_2 & -c_1 & 0 & 0 & 0 & 0 & d_1 \end{bmatrix}.$$

Another factored form is the *parallel* structure. This structure is obtained from a partial-fraction expansion of (2.18):

$$H(z) = \frac{e_1 z^{-1} + e_2 z^{-2}}{1 + c_1 z^{-1} + c_2 z^{-2}} + \frac{e_3 z^{-1} + e_4 z^{-2}}{1 + c_3 z^{-1} + c_4 z^{-2}} + \frac{e_5 z^{-1} + e_6 z^{-2}}{1 + c_5 z^{-1} + c_6 z^{-2}}.$$

$$(3.23)$$

Again, using the direct form II for each individual section results in the compensator structure of figure 3.9, which has 2 precedence levels, 12 coefficients, 7 delays (canonic), and 3 scaling multipliers:

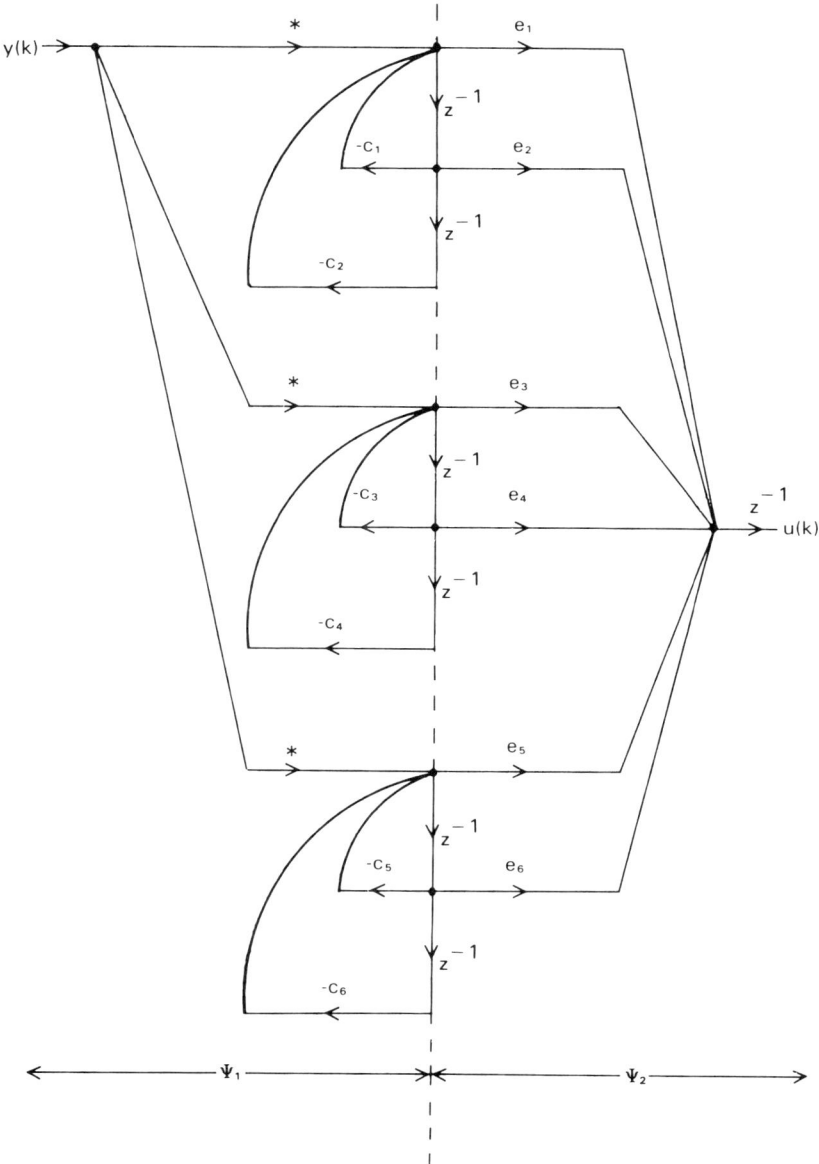

Figure **3.9** Parallel structure (direct form II).

$$
\Psi_2 =
\begin{bmatrix}
1 & 0 & 0 & 0 & 0 & 0 \\
0 & 1 & 0 & 0 & 0 & 0 \\
0 & 0 & 1 & 0 & 0 & 0 \\
0 & 0 & 0 & 1 & 0 & 0 \\
0 & 0 & 0 & 0 & 1 & 0 \\
0 & 0 & 0 & 0 & 0 & 1 \\
e_2 & e_1 & e_4 & e_3 & e_6 & e_5
\end{bmatrix},
$$

$$(3.24)$$

$$
\Psi_1 =
\begin{bmatrix}
0 & 1 & 0 & 0 & 0 & 0 & 0 & 0 \\
-c_2 & -c_1 & 0 & 0 & 0 & 0 & 0 & 1* \\
0 & 0 & 0 & 1 & 0 & 0 & 0 & 0 \\
0 & 0 & -c_4 & -c_3 & 0 & 0 & 0 & 1* \\
0 & 0 & 0 & 0 & 0 & 1 & 0 & 0 \\
0 & 0 & 0 & 0 & -c_6 & -c_5 & 0 & 1*
\end{bmatrix},
$$

The representation in (3.24) can be used for several structures since the real poles, if any, must still be grouped into sections. (The section-ordering and zero-pairing issues of the cascade disappear since all sections are in parallel, and the partial-fraction expansion gives no control over the zero locations.) Also, different types of second-order section structures are possible.

A structure that appears on the surface to be more natural for the LQG problem arises when we seek to implement the transfer function (2.17) directly with the parameters (coefficients) of equations (2.14):

$$
\Psi_3 \Psi_2 \Psi_1 =
\begin{bmatrix}
I_6 \\
\hline
-G
\end{bmatrix}
\begin{bmatrix}
\Phi & \Gamma & K
\end{bmatrix}
\begin{bmatrix}
I_6 & 0 \\
\hline
0 & \\
\hline
& -L & I_2
\end{bmatrix},
\qquad (3.25)
$$

where I_6 represents a 6×6 identity matrix. In general this structure (termed the *simple* form) has three precedence levels, is canonic in delays, and has up to $n(n + 4)$ coefficients, depending on the entries in Φ, Γ, L, K, and G. For a sixth-order LQG system, this structure after scaling could have up to 60 coefficients. This number of multiplies is quite excessive, compared to any commonly used filter structure. However, this compensator structure (or the similar structure based on the Ψ_∞ of the simple form) is often used for steady-state LQG control applications, more or less by default.

Another broad class of structures includes all the structures whose modified state-space representations have just one precedence level matrix. These structures could be called *state-space* structures, since the coefficients and the computations involved can be described using state-space notation. Some of these can be generated from the direct form II, cascade, parallel, and simple forms just by multiplying the various Ψ_i matrices together (using infinite-precision coefficient values) to produce Ψ_∞ and using the result as a structure. The standard observable, standard controllable, and Jordan forms [38] well-known to the control and estimation field also correspond to simple one-level structures [15, 31]. One could envision such structures being useful for two reasons. First, their performance may be superior to certain multiple-level structures, whether or not they have more coefficients. Second, a one-precedence-level structure allows a faster system sampling rate than a multiple-level structure (see chapter 4), and thus potentially better performance. An interesting type of one-level filter structure is the *minimum roundoff noise* structure of Mullis and Roberts [18, 39, 40], and Hwang [41]. Given no constraints on the coefficients of a one-level delay-canonic filter structure, they have derived a technique for computing the coefficient values producing minimum roundoff noise at the filter output. Unfortunately, this filter structure requires $(n + 1)^2$ coefficients. To avoid this problem, the authors have also presented *block optimal* filter structures, which are cascade or parallel forms composed of minimum noise second-order sections (see also Jackson, Lindgren, and Kim [42]). For a block optimal structure, only $4n + 1$ coefficients are required. One of the objectives of chapter 5 will be to extend the ideas of Mullis and Roberts to derive minimum roundoff noise *compensator* structures.

Using $f_1, \ldots, f_m$ as the coefficients, a sixth-order block optimal parallel compensator structure would have the following modified state-space representation:

$$\Psi_1 = \begin{bmatrix} f_1 & f_2 & 0 & 0 & 0 & 0 & 0 & f_3 \\ f_4 & f_5 & 0 & 0 & 0 & 0 & 0 & f_6 \\ 0 & 0 & f_7 & f_8 & 0 & 0 & 0 & f_9 \\ 0 & 0 & f_{10} & f_{11} & 0 & 0 & 0 & f_{12} \\ 0 & 0 & 0 & 0 & f_{13} & f_{14} & 0 & f_{15} \\ 0 & 0 & 0 & 0 & f_{16} & f_{17} & 0 & f_{18} \\ f_{19} & f_{20} & f_{21} & f_{22} & f_{23} & f_{24} & 0 & f_{25} \end{bmatrix}. \tag{3.26}$$

Note that the pole-zero pairing issue must still be addressed, as with any parallel form. No additional scaling multipliers are required in (3.26). As with any cascade, a block optimal cascade compensator structure would have the disadvantage of having multiple precedence levels—n_s in this case. (Recall that the parallel block optimal structure requires only one precedence level.)

Besides the direct form and general state-space forms, there exist other filter structures not derived from a factorization of the transfer function (2.18). Gray and Markel [43] have presented several ladder and lattice forms that are delay-canonic. Another set of ladder filters [44], also delay-canonic, results from *continued-fraction* expansions of (2.18). A ladder structure that has received a great deal of attention in the filtering literature is the wave digital filter [45–47]. This filter structure is based on analog LC ladder filters and directly results from a consideration of the transmission-line equations of microwave filters. Line delay and the transmitted and reflected voltage waves become the sample delay T and the signal variables of the wave digital filter. Characteristics of this structure that derive from the passivity and losslessness of its analog counterpart [48] carry over to the wave digital filter and lead to the absence of limit cycles under specific sign-magnitude truncation arithmetic (and see [49]). The coefficient sensitivity of this structure has been shown to be comparatively low [46], and under certain additional constraints [50] it will also be low-noise. Additional improvements have been introduced to reduce the number of multiplies [51] and the number of delays [52]. Meerkötter and Wegener [53] have developed a second-order wave digital filter section that can be the building block of a cascade or parallel form. This section has four multiplies and two sign-magnitude truncation quantizers but requires five additional scalers (as opposed to the one or two scalers of most sections). As with many of the digital filter structures, ladder-type structures could easily be adapted for compensators by manipulating the filter structure in such a way that the output node becomes a state node.

Finally, a general class of *optimal* structures exists. Chan [17] has described a technique for filters by which, through the use of the transformations in (3.14) and (3.15), a scalar function of the structure parameters can be minimized. More important, the method will hold almost any set of Ψ_i entries constant, as desired. Thus we can control the number of coefficients in the structure and their locations while minimizing roundoff noise, coefficient quantization effects, or some combination of the two. Chapter 8 will adapt this useful technique to the optimization of compensator structures and present an example of the constrained minimization of compensator roundoff noise effects.

This discussion of compensator structures was not intended to present an exhaustive list of possible structures, but only a representative selection. (For example, *transpose* configurations [31,33] were not considered.) The analyses in chapters 5–7 compare some of these compensator structures with respect to their finite wordlength properties. The overall aim is to provide the reader with a basic grasp of the various structures and of the different criteria for choosing among the different classes of structures, given control and estimation applications.

3.4 Summary

Beyond a presentation of the more common types of compensator structures, the main point of this chapter was the introduction of the modified state-space representation. This representation exactly reflects the computations that determine the performance of a compensator structure when implemented with finite wordlengths, and also the order in which these computations must occur. This representation, unlike the form introduced by Chan [17] that is useful for digital filters, will explicitly include all the inherent delays necessary to complete the operations within the compensator structure. Finally, as with the Chan form, it is possible to apply simple transformations to this representation in order to synthesize a compensator structure with superior finite-wordlength performance.

4

Architectural Issues: Serialism, Parallelism, and Pipelining

This chapter will examine the architectural issues involved in the implementation of digital feedback compensators. We shall show that the basic concepts of serialism and parallelism as they apply to digital filter structures represented in Chan's notation extend without modification to digital compensator structures represented in the modified state-space notation. However, the same cannot be said concerning the application of pipelining techniques to compensators. In fact, pipelining in control systems brings out another important issue: the interaction between the ideal design procedure described in chapter 2 and the implementation of the resulting compensator.

Perhaps the most basic issue in any consideration of digital system architecture involves the concepts of serialism and parallelism [33,54,55]. Essentially, this notion involves the degree to which processes, or operations, in the system run in sequence (serially) and the degree to which they execute in parallel (concurrently). At one extreme, *any* system can be implemented with a completely serial architecture, executing all its processes one at a time. This procedure requires the minimum number of actual hardware modules and the maximum amount of processing time for completion of the system task. On the other hand, any system can also be implemented with a maximally parallel architecture, having as many concurrent processes as possible. Such a design requires the maximal amount of hardware, but completes the overall system task in minimum time. Thus the serialism-parallelism trade-off is another example of the frequently encountered space-time trade-off [54].

There is an important asymmetry implicit in the exploitation of serialism and parallelism. It is always possible to execute processes one at a time (totally serially). However, it is *not* always possible to execute them all at once (in a totally parallel manner). There is a minimum amount of serialism required. Consider the simple model in figure 4.1 of a three-process system, with data cells [54] for input and output. (The computation of $a^2 + b^2$ would fit this model, for example, with processes $P1$ and $P2$ representing squarer modules, and process $P3$ representing an adder.) Assume that each of the three processes requires t seconds for completion

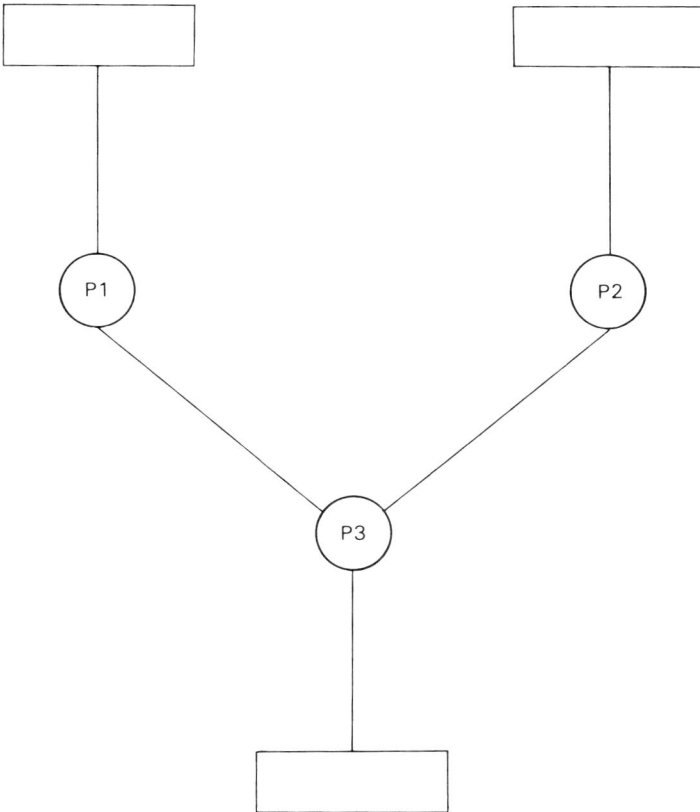

Figure **4.1** Three-process system.

(given specific hardware modules) and that each process executes as soon as all of its inputs are valid. Given a general-purpose computing module, then clearly a serial architecture that would require $3t$ seconds to complete the overall task is possible. On the other hand, figure 4.1 clearly shows that processes $P1$ and $P2$ must be finished *before* process $P3$ can begin. Consequently, only processes $P1$ and $P2$ can operate in parallel. For such an architecture, two hardware modules would be required, and the total computation time would be reduced to $2t$ seconds. The totally parallel architecture (total time t with three hardware modules) is not possible for the system of figure 4.1.

Under certain conditions, this "speed barrier" can be broken through the use of pipelining [33,54]. If the original objective of the system is to perform a task *repeatedly* (As soon as the present task is completed, a new task begins.), then pipelining could realize an effective throughput rate equal to (or at least closer to) that of a totally parallel architecture. Reconsider figure 4.1. Suppose that we assign a separate hardware module to each process and use the maximally parallel $2t$-second architecture described above. Also assume that the input and output data cells are implemented by storage registers clocked at the sampling rate $1/(2t)$. Now let us examine any $2t$-second interval. During the first t seconds, module 3 (for executing process $P3$) will be idle since its inputs are not yet valid. During the last t seconds, module 3 will be active and modules 1 and 2 will be idle. The total $2t$-second time from a task initiation until its completion cannot be reduced without faster hardware modules. However, the idle modules can be put to use by pipelining the processes. While module 3 is active and modules 1 and 2 otherwise idle, the next task can proceed and use modules 1 and 2. The net result (in this example) is a doubling of the throughput rate (task completions per second) from $1/(2t)$ to $1/t$. It must be stressed here that any given task still takes $2t$ seconds from start to finish; however, successive task completions occur at t-second intervals. In terms of the hardware required, the pipeline would be effected by adding two clocked registers to buffer the intermediate results from modules 1 and 2, and of course by doubling the clock rate. Figure 4.2 shows two ways of viewing the pipelined case for this example. Basically, the pipeline splits a larger task not implementable in a totally parallel architecture into smaller sequential subtasks, each of which can be implemented in a totally parallel fashion (figure 4.2a). An equivalent viewpoint (figure 4.2b) considers pipelining to be represented

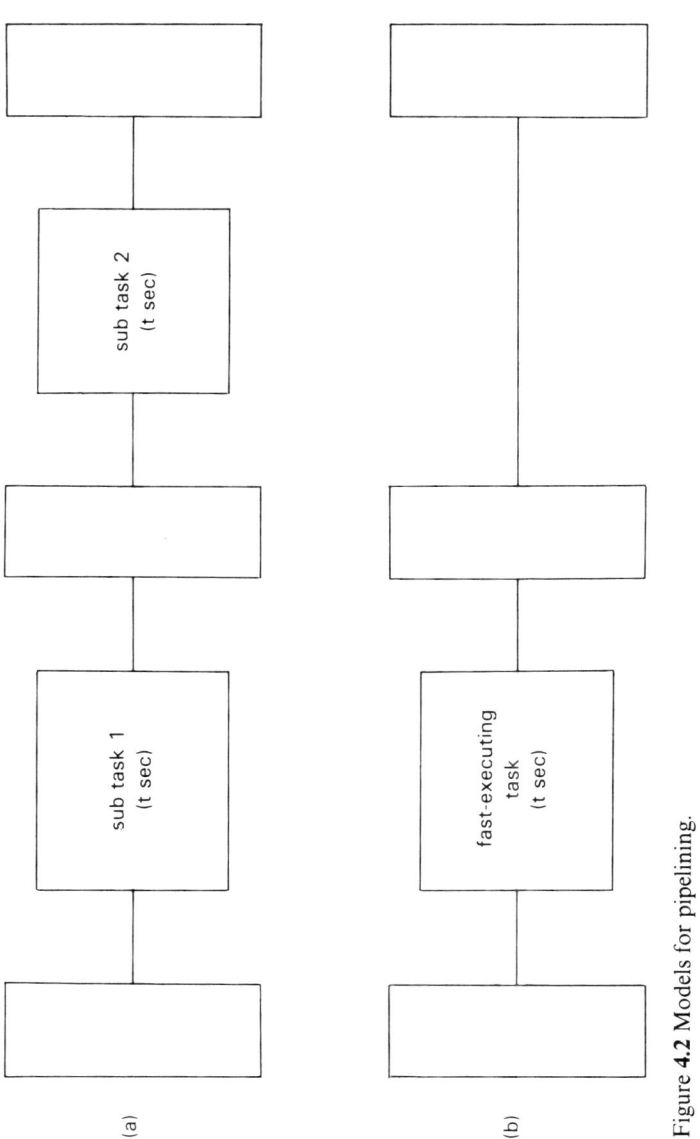

Figure **4.2** Models for pipelining.

by a faster-executing task coupled with some series delay (inherent in the clocked registers).

An important application of pipelining is in the implementation of digital filter structures [33, 56]. In such a case, the system task corresponds to the generation of a filtered output value from an input sample, and the individual processes correspond to the hardware digital multiplications, quantizations, and additions that exist in the particular structure implemented. (Ignore A/D and D/A operations for now.) Figure 4.3a shows a two-pole digital filter with input y and output u. As shown, the unit delay z^{-1} can be implemented as a clocked storage register. Thus all the arithmetic and quantization operations have one sampling period in which to be completed. Let us assume that quantization takes place after every multiplication. Then computing the signal $u(k + 1)$ at node A in figure 4.3a requires three multiplications and an addition. Also assume that we have selected an architecture for which the multiplications involving b_1 and b_2 operate in parallel; then the addition occurs, and finally the multiplication by a_1. Using three hardware multipliers instead of two, and assuming negligible add time, the multiply operations can be pipelined and the sampling rate doubled. The new configuration could be implemented with just one additional storage register, represented in figure 4.3b as an additional unit delay. However, this new signal-flow graph is not *node-minimal* since it contains two states that are *exactly* equivalent. Removal of one of these states produces the node-minimal signal-flow graph shown in figure 4.3c. Thus the pipelined structure of figure 4.3c has the same number of unit delays (storage registers) as the original structure in figure 4.3a. For this particular example, pipelining did not require the use of more unit delays. This would not be true in general. Note that each z^{-1} in figures 4.3b and 4.3c represents only half the delay time of those in figure 4.3a if the sampling rate is doubled (as made possible by pipelining).

From the example of figure 4.3, it is clear that pipelining ties in closely with the digital filter notion of precedence. Specifically, let us consider *node precedence*, that is, the precedence relations involved in the addition, multiplication, and quantization operations needed to compute the node signals. In this case the modified state-space representation (see chapter 3) is very convenient since it explicitly shows the number of precedence levels involved. If a structure represented in this notation has only one precedence level, then it can have a totally parallel architecture (parallel

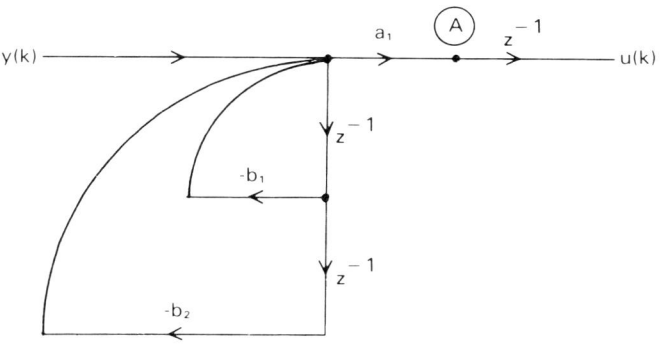

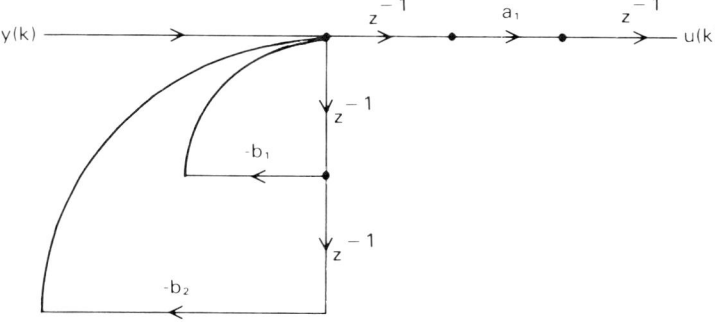

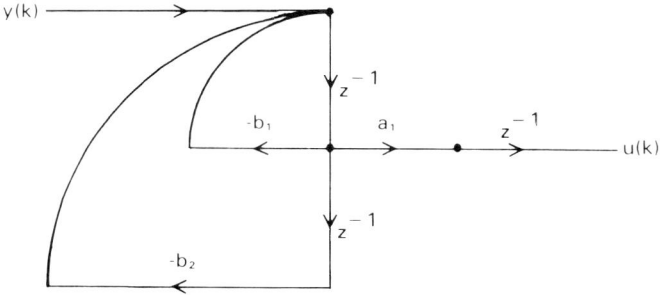

Figure **4.3** Pipelining a simple digital filter: (a) sample filter structure; (b) pipelined structure; (c) node-minimal pipelined structure.

in terms of the multiply-add computations involved in each precedence level). If more than one such level is required, no totally parallel architecture is possible, and the number of levels q will equal the minimium degree of serialism required. Pipelining, if applicable, would actually *change* the structure by inserting unit delays so that a new structure (one with fewer levels and thus a faster sample clock rate) is formed. The piplined structure would have the same transfer function as the original nonpinelined structure, except for some series delay, and would probably have more state nodes. Series delay is of little consequence in most digital filtering applications. Thus a two-level structure can be designed for a sampling period of $t/2$, even though the calculations require t seconds, since pipelinging (given a two-level structure) will fit the calculations into a $t/2$ slot at the expense only of a series delay of $t/2$ seconds. Equations (4.1)–(4.4) show the modified state-space representations and transfer functions of the nonpipelined (sampling period t) and pipelined (sampling period $t/2$) filters of figures 4.3a and 4.3c, respectively:

$$\Psi_2\Psi_1 = \begin{bmatrix} 1 & 0 \\ 0 & 1 \\ 0 & a_1 \end{bmatrix} \begin{bmatrix} 0 & 1 & 0 & 0 \\ -b_2 & -b_1 & 0 & 1 \end{bmatrix}, \tag{4.1}$$

$$H_{np}(z) = \frac{a_1 z^{-1}}{1 + b_1 z^{-1} + b_2 z^{-2}}, \tag{4.2}$$

$$\Psi_1 = \begin{bmatrix} 0 & 1 & 0 & 0 \\ -b_2 & -b_1 & 0 & 1 \\ 0 & a_1 & 0 & 0 \end{bmatrix}, \tag{4.3}$$

$$H_p(z) = \frac{a_1 z^{-2}}{1 + b_1 z^{-1} + b_2 z^{-2}}. \tag{4.4}$$

Note the reduction from two levels to one level [see (4.1) and (4.3)], allowing the doubled sampling rate, and also the extra z^{-1} factor in the numerator of (4.4). The number of states in (4.3) remained at three since no additional storage registers were actually added to effect the pipeline.

Let us now consider pipelining as it applies to the multiply operations in a structure alone. Such a consideration will be valuable whenever the multiply time far exceeds the addition and quantization operation times in a structure, a situation that is not uncommon in microprocessor-based digital systems. We would now consider the multiplier module to be the only significant process in the compensator. Since we are neglecting all calculation times other than the multiply times, it is sufficient to know the precedence to the multiply operations *alone* in order to determine the architectures that are possible. Thus the node precedence evident from the different Ψ_i matrices of a modified state-space representation will not be adequate to describe the *multiplier precedence* relations. Such relations can be determined from the signal-flow graph or from an examination of the specific location of each multiplier coefficient in the Ψ_i matrices. In either case, the multipliers can be grouped into precedence *classes*. Frequently, the number of *multiplier precedence classes* and *node precedence levels* will be the same, as in figure 4.3, but even in that case the multiplier coefficients in class 1 (of highest multiplier precedence) and the multiplier coefficients in node precedence level 1 (the matrix Ψ_1) need not be identical. It *will* be true that all the multiplier coefficients in the matrix Ψ_1 will also be in multiplier precedence class 1. Furthermore, multiple-level structures often have fewer multiplier classes than node precedence levels.

As an example, consider the cascade structure of figure 3.7 and its modified state-space representation (3.21). Assume all scaling multipliers to be simple shifts (powers of two); thus they are not considered to be true coefficients requiring hardware multipliers. All the multiplications of coefficients by state node or input signals can occur immediately after each sampling instant and therefore fall in multiplier precedence class 1. Thus the $c_1, c_2, c_3, c_4, c_5, c_6, d_2, d_3, d_4, d_5$, and d_6 multiplies can operate in parallel given enough hardware multiplier modules. Only the d_1 multiplication lies in class 2; it must await the completion of the c_1 and c_2 multiplies. Of course, given the two classes and 12 multiplies, an optimal, that is, maximal, use of the hardware is made with only 6 hardware multipliers (assuming no pipelining). Five of the class 1 multiplies (but not c_1 or c_2) would be computed in the *second* multiply cycle with the d_1 multiply. Thus the cascade of figure 3.7 has two multiplier precedence classes, although it has four node precedence levels. Similarly, the cascade structure in figure 3.8 has only one multiplier precedence class assuming power-of-two scalers, although its modified state-space representation (3.22) shows three node precedence levels. If in fact general scalers are used in these two cascades, they will constitute multiplier coefficients, and the number of multiplier precedence classes and node precedence levels will be the same. No matter what type of scalers are used, the parallel structure of figure 3.9 has the same number of multiplier classes as it has node precedence levels; even so, the coefficients of multiplier class 1 (c_1, c_2, c_3, c_4, c_5, c_6, e_2, e_4, and e_6) are *not* simply the coefficients in Ψ_1. The coefficients e_1, e_3, and e_5 belong to multiplier class 2 because they must await the completion of the c_1–c_6 multiplies. This notion of multiplier precedence is more completely formulated in [33], but the basic conclusion is as follows: Although the modified state-space representation correctly describes the operations that must occur in computing the node values within a structure and has other useful properties (see chapter 3), the multiplier precedence relations (more easily seen directly from the signal-flow graph) are more significant for determining the possible hardware architectures when the multiply time is dominant.

4.1 Restrictions on Pipelining

Certain basic restrictions [33] must be observed when pipelining a complex structure. The first limitation in applying pipelining concerns parallel data paths within the structure. Whenever any portion of a system is pipelined to increase the sampling rate (which adds effective delay), all parts of the system that feedforward in parallel with the pipelined portion must receive equivalent actual delay in order to maintain the desired transfer function. In other words, the data flowing through the system must remain synchronized whether or not pipelining is applied. Consider the second-order digital filter of figure 4.4a, with coefficients r_0, r_1, r_2, $-c_1$, and $-c_2$. A direct pipelining of this structure by adding a unit delay preceding the r_0 multiplier, as done with figure 4.3a, will result in a very different transfer function than the original one. To preserve the transfer function desired, except for series delay, unit delays must also be inserted in the parallel feedforward branches r_1 and r_2. This new (one-level) structure appears in figure 4.4b but is not node-minimal. Figure 4.4c shows an equivalent node-minimal structure, requiring only one additional state instead of three. Its modified state-space representation is shown in (4.5):

$$
\Psi_1 = \begin{bmatrix} 0 & 1 & 0 & 0 & 0 \\ 0 & 0 & 1 & 0 & 0 \\ 0 & -c_2 & -c_1 & 0 & 1 \\ r_2 & r_1 & r_0 & 0 & 0 \end{bmatrix}. \tag{4.5}
$$

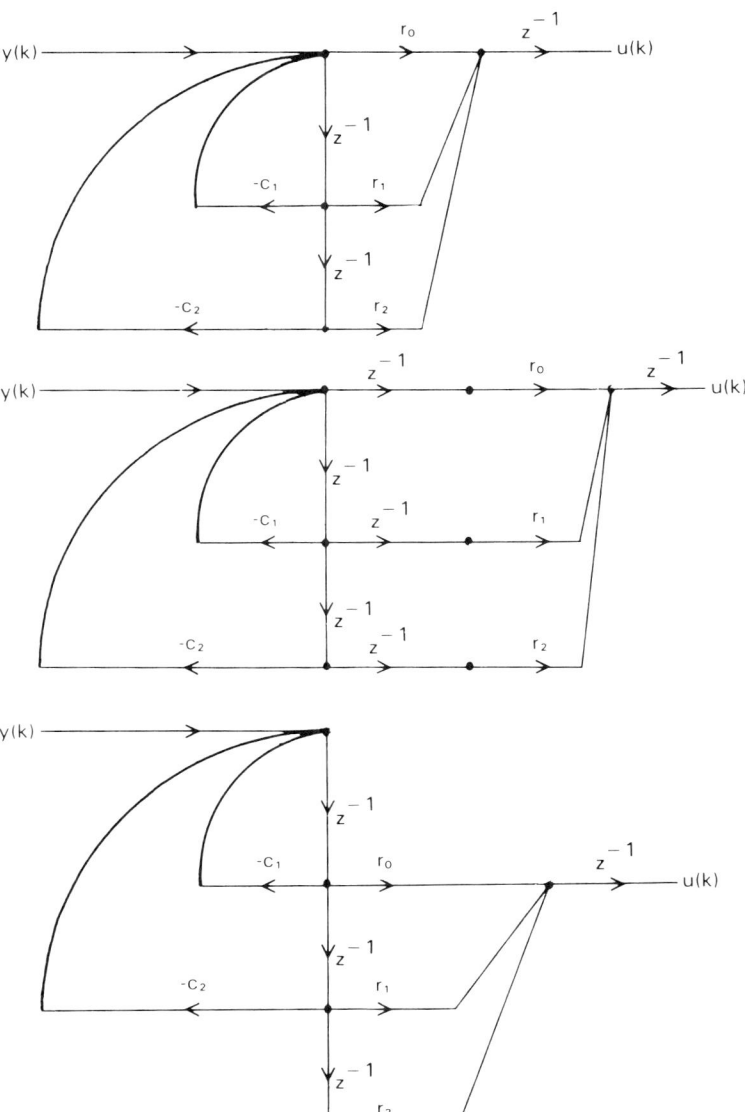

Figure **4.4** Pipelining and feedforward data paths: (a) filter structure; (b) pipelined structure, including feedforward paths; (c) node-minimal pipelined structure.

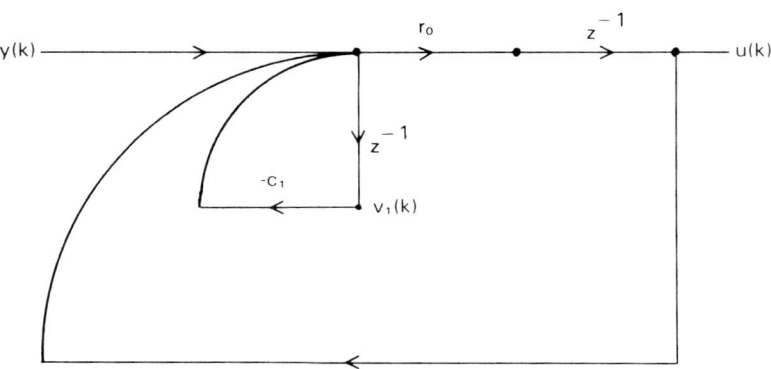

Figure **4.5** Filter with output feedback.

The second difficulty encountered in applying pipelining techniques involves feedback. Suppose there exists a series of operations that makes up part of a closed feedback loop within a structure. Pipelining these operations would result (as with the previous example) in a very different transfer function. Consider the filter of figure 4.5. Its transfer function and two-level modified state-space representation are shown in equations (4.6) and (4.7):

$$H(z) = \frac{r_0 z^{-1}}{1 + (c_1 - r_0)z^{-1}}, \tag{4.6}$$

$$\Psi_2 \Psi_1 = \begin{bmatrix} 1 \\ r_0 \end{bmatrix} \begin{bmatrix} -c_1 & 1 & 1 \end{bmatrix}. \tag{4.7}$$

If we pipeline by inserting a delay preceding r_0 (or by equivalently moving the r_0 branch to state node v_1), the modified state representation will indeed show only one level:

$$\Psi_1 = \begin{bmatrix} -c_1 & 1 & 1 \\ r_0 & 0 & 0 \end{bmatrix}. \tag{4.8}$$

However, the overall transfer function is now quite different:

$$H(z) = \frac{r_0 z^{-2}}{1 + c_1 z^{-1} - r_0 z^{-2}}. \tag{4.9}$$

Although part of the feedback loop has been "sped up" by pipelining, the delay introduced prevents the feedback term from being equivalently sped up. (The data is not synchronized.) Thus pipelining within a feedback loop is ordinarily avoided.

4.2 Pipelining Feedback Compensators

In the context of the control problem formulated in chapter 2, the ideas of serialism and parallelism apply unchanged to the implementation of digital controller architectures. However, since a global feedback loop exists around the entire compensator, that is, through the plant, pipelining seems to be out of the question, as shown in the example of figure 4.5. Suppose that we design an LQG compensator for a system with a sampling rate of $2/T$; the resulting compensator has two multiplier precedence levels, and the multiply time equals $T/2$. Pipelining would seem to be necessary unless we were willing to drop the sampling rate to $1/T$. Unfortunately, the series delay that would result from pipelining this compensator would introduce an unplanned pure time delay into the closed loop. The deleterious effects of pure time delay (linearly increasing negative phase shift) on the stability and phase margin of a feedback system are well known. Even if instability does not result, the performance as measured by the index J will be degraded and the qualitative dynamic performance will be compromised.

Fortunately, there is an approach to pipelining that can be effective for control systems. Consider the LQG system and compensator design technique described in chapter 2. Assume that for some original controller design, the sampling interval is not long enough to complete all the calculations involved in the compensator (which is the situation previously described). In principle, pipelining techniques could help, but unavoidable delay would be introduced. An effective use of pipelining simply means that we somehow include this unavoidable delay in the original design procedure. This aim can be realized through *state augmentation* [1]. Suppose that pipelining would allow an increase in the sampling rate by a factor of two, thus adding only a single series delay. If the plant is described at the doubled sampling rate $2/T$ by (4.10)

$$x(k + 1) = \Phi x(k) + \Gamma u(k) + w_1(k),$$
$$y(k) = Lx(k) + w_2(k),$$

$$(4.10)$$

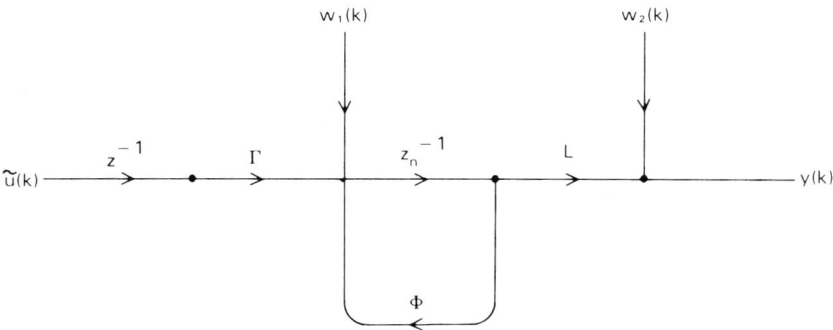

Figure **4.6** State augmentation for control system pipelining.

(recall that the matrix parameters above depend on T), then, preceding $u(k)$ with the series delay to form $\tilde{u}(k)$, the augmented plant can be modelled as follows (see figure 4.6):

$$\tilde{x}(k+1) = \begin{bmatrix} \Phi & \Gamma \\ 0 & 0 \end{bmatrix}\tilde{x}(k) + \begin{bmatrix} 0 \\ 1 \end{bmatrix}\tilde{u}(k) + \begin{bmatrix} w_1(k) \\ 0 \end{bmatrix},$$

$$y(k) = [L \ 0]\tilde{x}(k) + w_2(k), \tag{4.11}$$

where $\tilde{x}(k+1) = [x(k+1) \ u(k+1)]'$. For this augmented system, the weighting matrices Q and M in the expression for the performance index (2.6) must also be augmented, adding an all-zero row and column to Q, and a single zero element to M. The weighting parameter R will be the same as for the system (4.10). Now we must treat (4.11) as a new system and design an LQG compensator for it. Then *that* design can be pipelined, which introduces the inherent added delay shown in figure 4.6.

For this situation, two observations can be made. First, the Kalman filter portion of the LQG design for (4.11) will have what seems to be a difficulty due to the added delay—the numerical routines blow up. Common sense tells us, however, that there is no need to estimate $\tilde{x}_{n+1}(k) = u(k)$ since it is the *actual* plant input, which is known. Thus we need only estimate $\tilde{x}_1(k), \ldots, \tilde{x}_n(k)$, namely, the vector $x(k)$. That estimation problem has already been solved as the nth-order Kalman filter for (4.10), with gains $k_1, \ldots, k_n$. With these results, the optimal filtering gains for the augmented system (4.11) can be written

$$\tilde{k} = [k_1 \ k_2 \ \cdots \ k_n \ 0]'. \tag{4.12}$$

The $(n + 1)$th-order optimal regulator problem for (4.11) can be solved with no difficulty at all.

The second observation that we can make for this augmented-system pipelining technique involves the *consistency* of the design technique. A delay-canonic structure for the optimal LQG compensator for (4.11) will be of order $n + 2$ since (4.11) is of order $n + 1$, and *not* of order $n + 1$, as is the compensator structure for (4.10). Thus this approach to controller pipelining gives rise to a compensator of higher dimension (more poles), requiring more states (delay elements) and more coefficients. Along with this increase in order comes a more important point—the new higher-dimensional compensator structure must allow the same degree of pipelining as the original structure; otherwise, the whole controller pipelining design procedure is invalid, that is, inconsistent. This point is especially of concern when using structures whose number of precedence levels is a function of the number of compensator states (for example, the cascade forms). As an example, consider a second-order plant and a direct form II compensator structure, which requires three delays and two precedence levels. To exploit pipelining, we must augment the plant and redesign the compensator—its direct form II structure now requires four delays (states). There will still be only two (node or multiplier) precedence levels as before, so pipelining to double the sampling rate will work as planned. However, if we decide to use a *cascade* of two direct form sections (assume one second-order section, one first-order section, and scaling multipliers that are not in general equal to some power of 2, then the result is three precedence levels. Pipelining to allow the $2/T$ sampling rate will *not* now result in the effect of a single added unit delay as assumed, but will involve *two* series unit delays, making the design procedure invalid. In other words, if we implemented the pipeline as described, the system would not perform as expected; more delay would be present in the loop than had been accounted for in the design. Such problems can be avoided by a proper choice of structure.

There is one positive note associated with the increased dimensionality of the compensator, and it is related to the particular form of (4.12). Usually, an increase in dimension (number of states) by one involves at least two additional coefficient multipliers. (A fifth-order plant requires a compensator with at least 10 coefficients, a sixth-order plant requires one with 12 coefficients, and so forth—see figure 3.6.) However, by virtue of the zero entry in (4.12), the general form of the compensator transfer function for the augmented system is simpler:

$$H(z) = \frac{a_2 z^{-2} + a_3 z^{-3} + \cdots + a_{n+1} z^{-(n+1)}}{1 + b_1 z^{-1} + \cdots + b_{n+1} z^{-(n+1)}}. \qquad (4.13)$$

Comparing (4.13) to (2.18) shows a difference of only one coefficient— not two. This fact helps make the pipelining approach a bit more attractive, at least with certain structures (for example, any direct form and any cascade or parallel structure based on a direct form.)

One last general point should be mentioned. The application of any pipelining technique or the use of parallelism to increase the sampling rate is desirable only if it allows a decrease in the performance index J or whatever other gauge of system performance one accepts. However, not all systems have a performance measure that decreases (improves) monotonically with decreasing T [27]. Intuitively, any system with sharp resonances will lose controllability (implying a large J) when the sampling frequency is near a resonance. One must be aware of such cases. If such a case does not occur, then pipelining will reduce the performance index, although certainly not as much as the (nonimplementable) straight-forward LQG compensator design for sampling rate $2/T$ that adds no delay. Whether this pipelining approach is effective *enough* to warrant the higher-order compensator depends on the designer's particular application.

4.3 Controller I/O Pipelining

One common application of pipelining in a feedback environment involves the often time-consuming compensator input/output (I/O) operations, namely, the sampling and the A/D and D/A conversion operations. Let us assume that a structure with one multiplier precedence level [for example, the block optimal parallel structure of (3.26)] is chosen to implement a compensator and that a totally parallel architecture is used for the multipliers involved. Assuming negligible addition and quantization time, the compensator can be modelled as a three-process task (figure 4.7). With no pipelining the minimum sampling period T would equal $t_1 + t_2 + t_3$ seconds. Further assume that the slowest process is the multiply time and that $t_2 = t_1 + t_3 = T/2$. If we now pipeline these

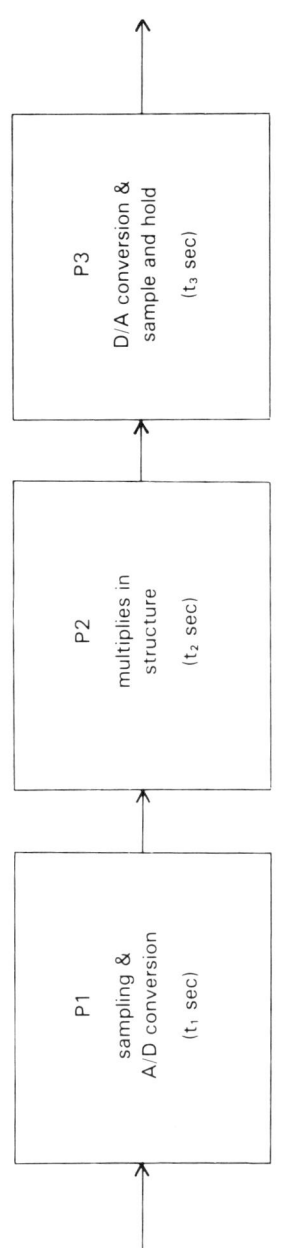

Figure **4.7** Three-process compensator model.

three processes, an increase by a factor two in throughput and sampling rate is possible. (Throughput rate is limited by the slowest process.)

Figure 4.8 diagrams the processes occurring in such an I/O pipelined compensator with increased sampling rate $2/T$. At each sample time, sampling and A/D conversion of a new y sample would begin. Then t_1 seconds later the structure multiplications could begin, overlapping the next sampling and A/D operation. (Note that the hardware multipliers will now be active 100% of the time.) We can represent this pipelined system as the designed compensator structure followed by a series unit delay resulting from the pipeline. Since part of this unit delay is involved in buffering the intermediate A/D results and the rest is involved in buffering the multiplier results from the structure, two hardware storage registers will be required for this example. However, their clock signals will be staggered since the three operations of figure 4.8 take different amounts of time to complete. Basically, these clock signals (all of period $2/T$) must be phased so that the results from each process are stored as soon as they are completed. Thus register 1 is clocked by sample pulses delayed by t_1 seconds, and register 2 is clocked by sample pulses delayed by $t_1 + t_2$ seconds. (This phasing is shown as fractional delay time in the simple example of figure 4.9.)

If we apply the design technique outlined in section 4.2 to produce a (pipelinable) compensator for this I/O case, the order of the compensator will of course be one greater than the nonpipelined design, implying at least one additional state and coefficient. No matter what the plant dimension may be, a block optimal parallel structure (or any state-space structure—see section 3.3) will have only one precedence level. Thus, I/O pipelining with such a one-level compensator structure results in a valid design procedure.

4.4 Compensator I/O Pipelining Examples

Four examples have been selected to illustrate what can occur with compensator (I/O) pipelining. Each example consists of four cases. Case 1 represents the plant discretized at a T-second sampling period with its corresponding LQG compensator (no pipeline). Case 2 represents the plant discretized at a $T/2$-second sampling period with *its* corresponding LQG compensator. This case does not include any pipelining, and thus is not physically implementable due to the short sampling interval. The

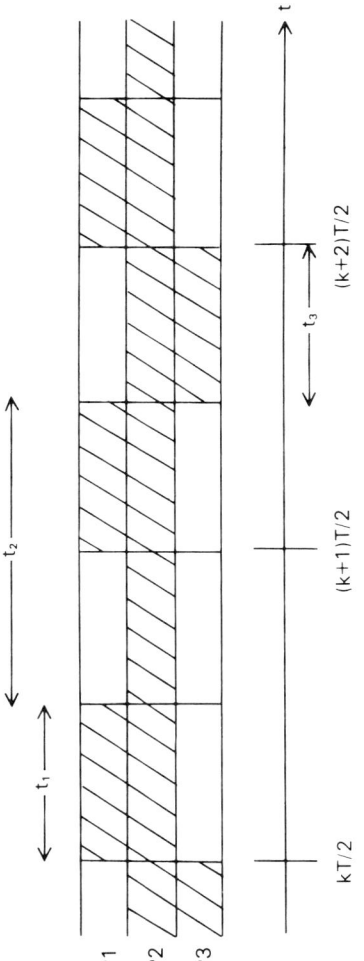

Figure **4.8** Process concurrency in I/O pipelined compensator.

performance index for this case constitutes an unreachable lower bound to the performance of the augmented-plant approach to pipelining (case 3). Case 4 (blind pipelining) results when the compensator designed for case 2 is pipelined in order to make it physically implementable. Thus the delay due to the pipeline is *ignored* in the pipelined design, usually resulting in a performance level that is worse than the nonpipelined level (and perhaps even in a system that is unstable). Assuming that J is a monotonic increasing function of T, we can expect that the different cases will rank, from highest J to the lowest, as follows: case 4, case 1, case 3, case 2. (It is possible but unlikely that case 4 could have a lower J value than case 1.) Remember, however, that case 2 is not implementable.

The simplest I/O pipelining example consists of a single-input, single-output, single-integrator plant:

$$\begin{aligned} \dot{x}[t] &= u[t] + w_1[t], \\ y[t] &= x[t] + w_2[t], \end{aligned} \tag{4.14}$$

where $T = 6$ seconds. Referring to chapter 2, equations (2.1)–(2.3), the parameters $\hat{Q}$ and $\hat{R}$ were both chosen to be 1 and the noise intensities Ξ_1 and Ξ_2 were selected to be 0.3 and 0.125. Figure 4.9 illustrates the discretized system and the form of the compensator before pipelining (case 1) and after pipelining through state augmentation and redesign (case 3). A one-level version of the direct form II structure (obtained from the Ψ_∞ matrix of the direct form II, as mentioned in section 3.3) is used for the compensator. Note the inclusion of the two fractional delays (registers) in figure 4.9b, as mentioned earlier in this section. The form of the system for case 2 would look the same as that in figure 4.9a; however, the gains of the branches would differ. For case 4, we need only add one series delay to the signal-flow graph of case 2.

Three other examples are also considered: a double-integrator plant, a two-state harmonic oscillator plant, and a sixth-order plant derived from the longitudinal dynamics of the F8 fighter aircraft (see chapter 5 and appendix A). The continuous-time parameters of the double-integrator system are

$$\dot{x}[t] = \begin{bmatrix} 0 & 1 \\ 0 & 0 \end{bmatrix} x[t] + \begin{bmatrix} 0 \\ 1 \end{bmatrix} u[t] + w_1[t],$$

$$y[t] = [1 \ \ 0] x[t] + w_2[t]. \tag{4.15}$$

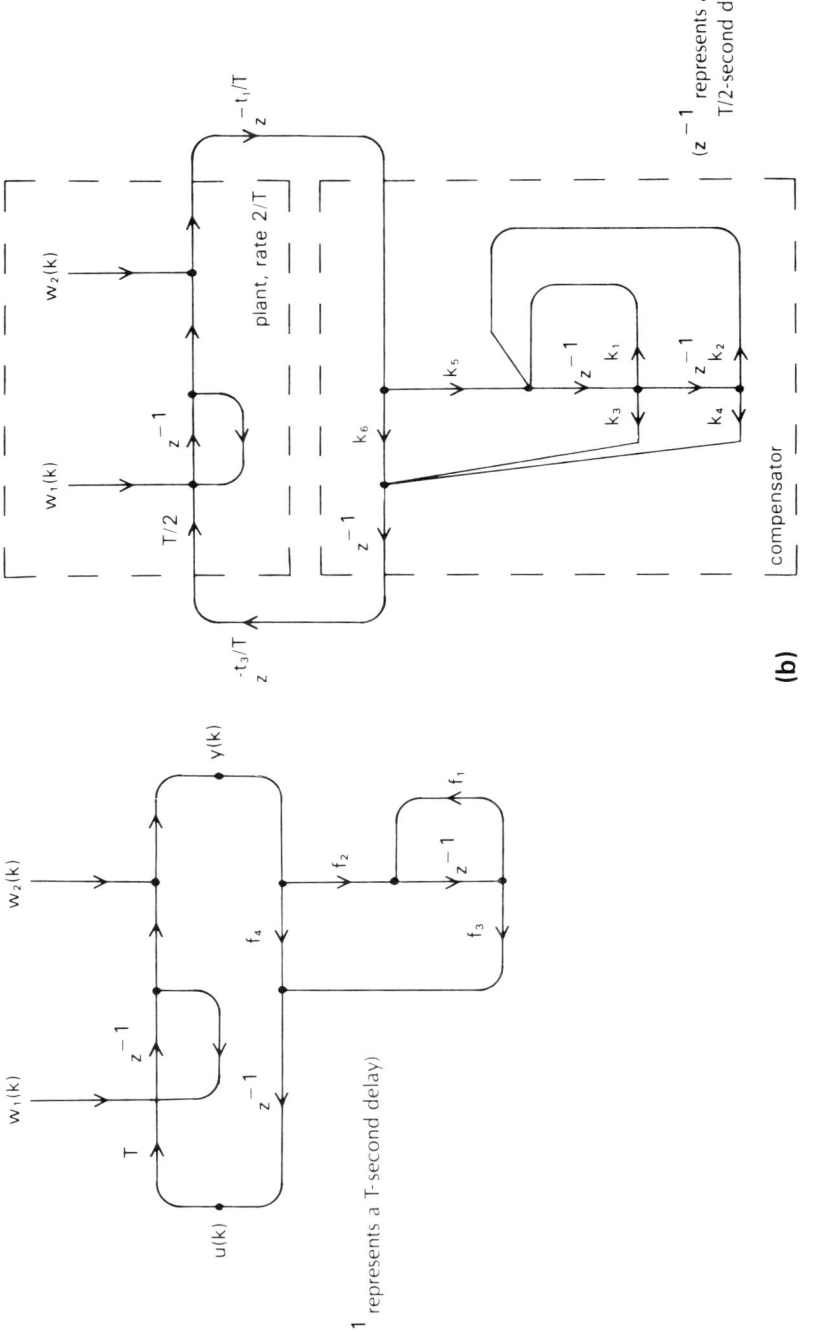

Figure 4.9 Compensator I/O pipelining, single-integrator plant: (a) rate $1/T$ system, $T = 6$ (case 1); (b) pipelined system, rate $2/T$ (case 3).

Table **4.1** Compensator I/O pipelining[a]

Example plant	T	Case 4	Case 1	Case 3	Case 2
Single integrator	6	(unstable)	2.42	2.05	1.34
Double integrator	6	(unstable)	328	179	53.2
Harmonic oscillator	6	(unstable)	32.7	12.9	9.72
6-State F8 plant	1	0.0038	0.00312	0.00282	0.00222

a. Key: case 1, rate $1/T$ system; case 2, rate $2/T$ system (not implementable); case 3, rate $2/T$ pipelined system designed via state augmentation; case 4, blind pipelining.

For this system the continuous-time parameter $\hat{Q}$ was a 2×2 identity matrix, $\hat{R}$ was 1, Ξ_1 was the diagonal 2×2 matrix diag(0.2, 0.3), and Ξ_2 was 0.125. For the harmonic oscillator, all the parameters were the same as for the double-integrator system, except for the A matrix:

$$A = \begin{bmatrix} 0 & 1 \\ -1 & 0 \end{bmatrix}. \tag{4.16}$$

The performance indices for all the various cases are shown in table 4.1. Under case 4 we see the consequences of pipelining and ignoring the delay incurred. Three of the example systems actually became unstable, and with the fourth the index J *increased*. As expected, all the case 2 indices were lower than in case 1, with case 3 lying between the two. To judge the effectiveness of the state-augmentation pipelining method of case 3, one must examine the degree of improvement in J relative to the possible improvement (the difference between cases 1 and 2). The best improvement shown was for the harmonic oscillator, which is no surprise since the oscillator's natural frequency of $1/(2\pi)$ radians/second is close to the unpipelined sampling rate $1/T$. The remaining three examples also showed significant improvement. Again, whether the pipelinable compensator (with one extra state and at least one extra coefficient) is to be used will depend on the particular level of performance desired and the penalty involved in complicating the hardware.

4.5 Summary

This chapter has investigated certain architectural issues associated with the implementation of digital feedback compensators. The introduction presented the notions of serialism, parallelism, and pipelining, and explained the hardware cost-execution time trade-off tied to these issues. The issues of serialism and parallelism were shown to involve the same considerations for digital compensators as for digital filters. Section 4.1 discussed the limitations of pipelining techniques, especially the one concerning pipelining in a closed loop (feedback). The extra delay incurred due to the use of pipelining had a deleterious effect on the performance of the feedback system. This problem made the consideration of pipelining for feedback compensators very different than in the case of digital filters. Section 4.2 developed a design technique based on state augmentation for dealing with the problem of control system pipelining. Finally, the last two sections treated a typical application of pipelining techniques to microprocessor-based control systems. For this application the compensator input, output, and multiply operations could be pipelined to realize a doubling in the system sampling rate. Four examples were presented to illustrate the technique.

5

Finite-Wordlength Effects: Quantization Noise

Perhaps the most important issues arising in the implementation of digital compensators involve the effects of finite precision. In order to reduce the cost of the controller, we wish to store the necessary node values and parameters as efficiently as possible, that is, with as few bits as possible. However, finite wordlengths lead to degraded performance. Therefore we need a methodology for selecting appropriate wordlengths and estimating performance degradation due to those finite wordlengths.

There are essentially four areas in which finite precision affects the implementation of the compensator. First, the ideal parameters of whatever structure we have chosen must be represented in a limited number of bits. This *coefficient quantization* issue is treated at length in chapter 6. Second, the A/D converter at the input of the compensator must translate an analog signal to a limited digital representation using n_{ad} bits. Thus quantization errors occur. The performance degradation due to these errors will depend on n_{ad}, but not (to first order) on the structure chosen. The third source of finite wordlength effects arises from the multiplication operations in the structure. Consider a compensator using fixed-point arithmetic in which the node signal values are represented with n_r fractional bits (thus these values are always less than one) and the coefficients are represented with one integer bit and n_c fractional bits. Every multiplication will then produce a product with $n_r + n_c$ fractional bits. This product must either be stored as a new node signal value, added to other products and the sum stored, or added to other products and passed on to another multiplier. In any case, quantization to n_f fractional bits is required, and errors are generated. Unlike the A/D quantization case, the effects of these errors *are* dependent on the particular structure used. With this same example, we can see a fourth case of finite wordlength problems: *overflow*. Clearly the product mentioned above could be greater than one (have an integer bit), or at least a sum of products less than one could be greater than one. In either case, an overflow occurs since the result must be stored as a node signal value less than one. However we choose to implement this overflow, whether by saturating or by some other nonlinear function (see chapter 7), errors will occur and performance will be affected.

The A/D and multiplier quantizations mentioned introduce two types of undesirable effects, roughly classifiable as *periodic* and *random*. The periodic effects (limit cycle oscillations) will be treated in chapter 7, along with the periodic effects of certain overflow nonlinearities. The random effects, *quantization noise*, are the subject of this chapter.

Before discussing techniques for analyzing quantization noise, the issue of *scaling* must be dealt with. Scaling essentially normalizes the coefficients of a structure so that overflows are prevented, or at least held to some low level, while simultaneously keeping the quantization effects to a minimum. Without this normalization, it is really not possible to compare the quantization noise effects of two different structures. Thus this chapter will also include the development of scaling procedures for compensators.

As mentioned in chapter 4, LQG compensators have an inherent measure of performance, the index J. This index reflects the weighted root-mean-square fluctuations of the plant states and controls. As we shall show, the effects of quantization noise can be treated as an increase in this index. It is not necessary to use this measure, since we could consider the output noise power, or some other noise metric, but the results can be extended to these other measures in a relatively straightforward manner. If the problem under consideration had been a Kalman filter, then the appropriate measure would have been the trace of the error covariance matrix. We have simply selected the index J as the most appropriate for the LQG setting.

In the discussions of scaling and quantization noise analysis, there are certain assumptions, or distinctions, that we shall make. First, the storage registers (and quantizers) within a structure may have different, *nonuniform*, wordlengths; such a structure can always perform better in terms of roundoff noise effects than the constrained case of uniform wordlengths [18]. However, by using uniform wordlengths, the hardware expense and complexity will be greatly reduced and the scaling and analysis procedures simplified. Typically, little potential performance is lost by such a restriction. Also, since the A/D converter is usually a separate piece of hardware, little affected by the remaining compensator hardware architecture and design, it need not be subject to this restriction. Consequently, A/D and internal wordlengths can, and typically do, differ. We shall assume that the signal variable registers are of uniform wordlength and that the A/D wordlength can be different from the internal compensator wordlength.

The second assumption, or distinction, concerns the placement of the structure's quantizers. On one hand, they can be inserted after *every* multiplication—ensuing adders would thus have to deal only with n_r-bit quantities. However, if we are willing to complicate the adders, we can place the quantizers just *after* the node additions. With this method, the adders would have to sum $(n_r + n_c)$-bit quantities, but fewer quantizers would be needed. This alternative trades off hardware complexity (double-precision versus single-precision adders) for quantization noise. (Fewer quantizers implies fewer noise sources, and thus less degradation in performance.) Both these options will be considered in this chapter.

The final distinction in discussing quantization noise is in the *type* of quantizer used. Commonly, the choice is between *rounding*, which selects the finite-precision word that is closest to the ideal value, and *truncating*, which simply drops the extra bits of precision. Truncation, and specifically sign-magnitude truncation, requires little or no extra hardware, and also has an advantage in terms of the resulting (reduced) number of possible limit cycle oscillations. However, rounding can be shown to have reduced quantization noise effects, and the extra hardware it requires is not very complex. In addition, roundoff effects are more easily analyzed. Consequently, this chapter will focus primarily on roundoff quantization. In chapter 7 we shall consider other quantization methods that provide advantages in terms of limit cycle behavior, that is, fewer limit cycles or limit cycles of smaller amplitude.

This chapter is organized as follows. Section 5.1 will discuss the major issue of dynamic range and scaling as applied to digital filters. Section 5.2

will adapt these ideas for digital control compensator scaling. This adaptation will have to consider the entire closed-loop system in determining the appropriate scaling for compensators. Set-point LQG configurations and their implications as regards the scaling issue will also be discussed. Section 5.3 will describe the roundoff and sign-magnitude truncation quantization characteristics and present models that can be used in analyzing their effects. Quantization noise analysis methods using the roundoff model will be treated in section 5.4. Section 5.5 will describe the minimum roundoff noise filter structures introduced by Mullis and Roberts [18, 39, 40] and Hwang [41] and then will adapt these results to derive minimum roundoff noise compensator structures. Finally, section 5.6 will demonstrate the procedures developed in chapter 5 for compensators by applying them to 10 candidate structures for implementing a specific control system.

5.1 Dynamic Range Constraints

It is not meaningful to discuss quantization noise effects (which are fundamentally related to the least significant bit of the node signal value) without also considering the dynamic range of the signals within the structure. Let us assume that the overall objective is to minimize the total number of bits necessary for the uniform-length fixed-point digital words. Choosing a specific structure based on its required least significant bit size (quantization step size) is of little value unless the fixed-point words can represent the full dynamic range of the node signals while keeping overflows to a minimum. Thus we must maximize signal-to-noise ratio without incurring overflow. These aims can be accomplished through *scaling*. By scaling the coefficients of a structure we can reduce the overall dynamic range of the signals within the structure and also normalize the maximum signal size (the overflow level) at each node. Once a structure is scaled, we *can* use the quantization step size as a valid basis for comparison with other structures which have been scaled using the same scaling procedure. Note that scaling should not alter the type of structure nor its ideal transfer function.

Consider the second-order filter of figure 5.1a. This structure has three states, implying three storage registers. Clearly, if the $v_2(k + 1)$ node and the $u(k + 1)$ output node do not overflow, then none of the node signals will overflow, since the other nodes are simply delayed versions of these

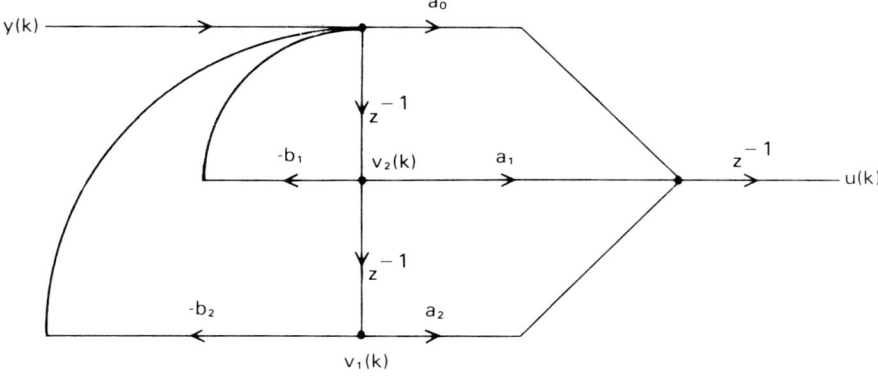

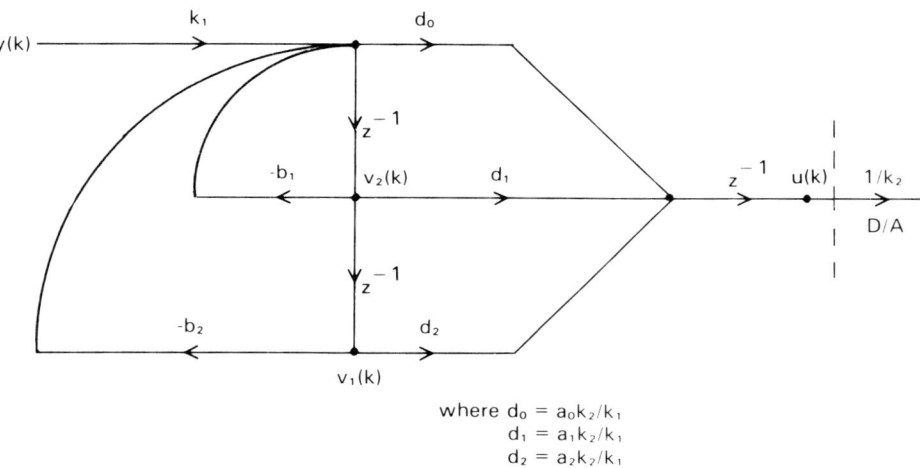

where $d_0 = a_0 k_2/k_1$
$d_1 = a_1 k_2/k_1$
$d_2 = a_2 k_2/k_1$

Figure **5.1** Scaling a second-order section: (a) unscaled; (b) scaled.

two. Thus scaling involves overflow constraints on these two nodes. Such constraints would be inequality constraints; that is, the signal magnitude must be *less* than the overflow level. Of course, too small a signal magnitude would result in higher quantization noise levels. Intuitively, we would like to alter the magnitudes of the signals at these two nodes just enough to prevent the occurrence of overflow, but without changing the filter transfer function. For example, to modify the signal magnitude at the $v_2(k + 1)$ node, the input unity coefficient must be multiplied by some factor k_1, and then to preserve the transfer function of the filter, the three coefficients a_0, a_1, and a_2 must be multiplied by $1/k_1$. Similarly, scaling the $u(k + 1)$ node involves multiplying a_0, a_1, and a_2 by another factor, k_2. The corresponding $1/k_2$ factor must then be absorbed by the output D/A converter to ensure an unchanged overall transfer function. The resulting scaled structure is shown in figure 5.1b.

An important choice must be made in selecting k_1 and k_2. Let us define *optimal* scaling to refer to the particular choice of scalers that satisfies the dynamic range constraints of the scaling procedure (inequality constraints) with equality. Thus, in general, such scalers will not be simple powers of two. For the example given, optimal scaling would result in a structure with six nontrivial multiplications, instead of five. Optimal scaling will nearly always be superior to nonoptimal scaling (which results when the scalers are constrained to be simple powers of two to simplify the hardware) in that the resulting structure will exhibit reduced quantization noise effects, even with the extra noise sources that can result from the additional scaling coefficients. Thus scaling introduces another trade-off between performance and hardware complexity.

Two basic approaches exist for determining a set of dynamic range constraints. The first is a deterministic norm-based method introduced by Jackson [57]. Let us define the L_p norm of a digital frequency-domain transform $H(z)$ as follows:

$$\|H\|_p = \left\{ \frac{1}{\omega_s} \int_0^{\omega_s} |H(e^{j\omega T})|^p \, d\omega \right\}^{1/p}, \tag{5.1}$$

where ω_s is the sampling frequency in radians per second. If $F_i(z)$ is defined to be the transfer function from the input to the ith node that must be scaled, then it can be shown [55] that

$$|g_i(k)| \leqslant \|F_i\|_p \|U\|_{p_0} \text{ for } \frac{1}{p} + \frac{1}{p_0} = 1, \qquad p, p_0 \geqslant 1, \quad \text{and all } k, \qquad (5.2)$$

where $g_i(k)$ is the signal at the ith node to be scaled and $U(z)$ is the z-transform of the filter input $u(k)$. Note that when this inequality is applied to $u(k)$ itself $[g_i(k) = u(k), F_i(z) = 1]$, we find that $|u(k)| \leqslant M_0$ if $\|u\|_{p_0} \leqslant M_0$ for *any* $p_0 \geqslant 1$.

Now let us return to the scaling issue for node i. Assume that the maximum signal magnitude possible in the filter without overflow is M_0. Further assume that $\|U\|_{p_0} \leqslant M_0$, and thus u never overflows (its magnitude is always $\leqslant M_0$, as mentioned). Then using (5.2), the node signal g_i will not overflow if

$$\|F_i\|_p \leqslant 1 \qquad \text{for all} \quad i. \qquad (5.3)$$

This scaling rule, which is called L_p scaling, must be satisfied at every node in the filter structure. Satisfying this rule with equality corresponds to *optimal* scaling. For the example of figure 5.1, the scaling multipliers k_1 and k_2 would be chosen to satisfy (5.3) for $i = 1$ and $i = 2$.

The L_p scaling rule just described still allows some degree of freedom even with optimal scaling, namely, the choice of p_0 and p. If we know that the L_1 norm of the input (the average value of its z-transform magnitude) is less than M_0, then we can only use L_∞ scaling. However, if the L_2 norm of the input is less than M_0, we can use either L_∞ or L_2 scaling. This follows from the monotonicity of the L_p norm [58]: $L_i \leqslant L_j$ if $i < j$. In this case ($p_0 = 2$) we would select L_2 scaling over the L_∞ choice. They both would eliminate overflow, but the L_2 rule would result in lower quantization noise effects. In other words, the smaller the value of p, the less conservative the scaling procedure will be. Thus the more we can restrict the possible filter input signals (the larger we can make p_0), the better the scaling will be in terms of the resulting noise levels.

A related deterministic scaling method has been described by Hwang [58]. This method is based on the time-domain l_p norm, which can be defined for a semi-infinite sequence $\{h_i(k)\}$ as

$$\|h_i\|_p = \left(\sum_{k=0}^{\infty} |h_i(k)|^p \right)^{1/p}. \qquad (5.4)$$

Note that the l_2 norm corresponds to the square root of the summed squared value of the sequence h_i, and by Parseval's relation equals the

L_2 norm of $h_i(z)$, the z-transform of h_i. We can now write the time-domain counterpart of (5.2) as follows:

$$|g_i(k)| \leqslant \|f_i\|_p \|u\|_{p_0} \qquad \text{for } \frac{1}{p} + \frac{1}{p_0} = 1, \quad p, p_0 \geqslant 1, \tag{5.5}$$

where g_i is the signal at the ith node to be scaled (as before), $f_i(k)$ is the impulse response of node i at time k, and $u(k)$ is the filter input. The following scaling law results: If M_0 is the maximum signal magnitude allowed in the filter and $\|u\|_{p_0} \leqslant M_0$, then

$$\|f_i\|_p \leqslant 1 \qquad \text{for all } i \tag{5.6}$$

guarantees no overflow.

In order to compare L_p and l_p scaling methods, we must examine the relation between the L_p and l_p norms [58]:

$$\|u\|_\infty \leqslant \|U\|_1 \leqslant \|U\|_2 = \|u\|_2 \leqslant \|U\|_\infty \leqslant \|u\|_1. \tag{5.7}$$

Given (5.7), we can determine how conservative any given scaling rule is as compared to all the other scaling rules. From (5.7), we know that if the input satisfies the constraint $\|U\|_1 \leqslant M_0$, then it must also satisfy $\|u\|_\infty \leqslant M_0$ (the reverse is not true). Thus, as far as the type of input signal is concerned, knowing that the L_1 norm of the input is less than M_0 is *more* restrictive than knowing that its l_∞ norm is less than M_0. (The l_∞ norm is simply the maximum magnitude value of a sequence, so knowing that the l_∞ norm is $\leqslant M_0$ only tells us what has already been assumed for the purposes of scaling.) We can generalize this statement to the entire list in (5.7). Since a less restrictive input corresponds to a more conservative scaling, we can use (5.7) to determine how any scaling method compares to any other. Thus the least conservative scaling is l_∞ scaling, and the most is l_1 scaling. The actual scaling method selected will depend on what information is known about the filter input signal or its transform. It is important to note that whenever the input has a nonzero constant (DC) component, we must use fairly conservative scaling, since the only finite-valued norms of the input would be the l_∞ and L_1 norms.

The second approach to establishing dynamic range constraints and choosing scaling multipliers is stochastic [18, 39, 41]. Suppose that we treat the filter input as a random process and consider the *probability of overflow* at each node rather than trying to prevent overflow completely. In terms of minimizing overflows, the best approach is now to scale in such a way that we equalize the probability of overflow at each node. As an example, let us assume that the maximum signal level without overflow is M_0 and that the input is a zero-mean white Gaussian random process with standard deviation $M_0/3$. The probability of overflow at the input A/D is then 0.003. The deviation of the signal at node i can be written

$$\sigma_i = \frac{M_0}{3}\left(\sum_{k=0}^{\infty} [f_i(k)]^2 \right)^{1/2}. \tag{5.8}$$

However, this quantity is just the l_2 norm of $f_i(k)$ multiplied by the input deviation. Thus each node will have the same probability of overflow only if $\| f_i \|_2 = 1$ for all i. Thus stochastic scaling to force equal overflow probability will be equivalent to l_2 or L_2 deterministic (optimal) scaling.

In terms of a state-space structure as discussed in Mullis and Roberts [18, 39], scaling corresponds to a *diagonal* similarity transformation of the unscaled structure. In the more general context of Chan's notation or the modified state-space representation introduced in chapter 3, scaling can be described by a set of diagonal scaling matrices S_i. In the remainder of this section we shall review Chan's procedure [17] for scaling filters as it would be developed using the modified state-space notation of chapter 3. (Thus the output node of the filter will be a state node.) Section 5.2 will then extend scaling ideas to the control setting.

In the context of the modified state-space representation (3.17), let us now examine stochastic l_2 scaling for filters (adapted from [17]). Let us partition $\Psi_{\infty} = \Psi_q \cdots \Psi_1$ (defined in section 3.3) as follows:

$$\Psi_{\infty} = [\Psi_{11} \ \Psi_{12}], \tag{5.9}$$

where Ψ_{11} is $(n + 1) \times (n + 1)$ and Ψ_{12} is $(n + 1) \times 1$. Assume the structure coefficients to be of infinite precision. (This assumption will hold throughout the remainder of this chapter. The case of quantization noise assuming finite precision coefficients is discussed in section 6.6.) Thus the states, input, and output of the filter can be described with the following state space of order $n + 1$:

$$\begin{bmatrix} v(k + 1) \\ u(k + 1) \end{bmatrix} = \Psi_{11} \begin{bmatrix} v(k) \\ u(k) \end{bmatrix} + \Psi_{12} y(k),$$

$$u(k) = \begin{bmatrix} 0 & 0 & 0 & \cdots & 0 & 1 \end{bmatrix} \begin{bmatrix} v(k) \\ u(k) \end{bmatrix}. \tag{5.10}$$

For this system of equations, the state covariance matrix V can be written

$$V = \overline{\begin{bmatrix} v(k) \\ u(k) \end{bmatrix} [v'(k)\ u(k)]} = \varepsilon \sum_{j=0}^{\infty} (\Psi_{11}^{j-1} \Psi_{12})(\Psi_{11}^{j-1} \Psi_{12})'. \tag{5.11}$$

Let us define the matrix K_q to be V/ε. We can write a Lyapunov equation that is equivalent to (5.11) and usually easier to evaluate:

$$\Psi_{11} K_q \Psi_{11}' + \Psi_{12} \Psi_{12}' = K_q. \tag{5.12}$$

The diagonal elements of K_q represent the gains from the input variance to the state node variances. Now let us consider the intermediate node variances, assuming that the structure is multilevel. Since the intermediate nodes are related to the state nodes via the precedence level matrices $\Psi_1, \ldots, \Psi_{q-1}$, we can compute a set of matrices K_i whose diagonal elements are the gains from the input variance to the variances of the intermediate node vector r_i:

$$K_i = \Psi_i \Psi_{i-1} \cdots \Psi_1 \begin{bmatrix} K_q & 0 \\ 0 & 1 \end{bmatrix} \Psi_1' \cdots \Psi_{i-1}' \Psi_i' \qquad \text{for } i = 1, \ldots, q-1. \tag{5.13}$$

Stochastic scaling (l_2 scaling), which equalizes the probability of overflow at all the nodes in the structure including the input, can be realized by forcing all the diagonal entries of the K_i matrices to unity. Thus all the node variances will be the same as the input variance. This scaling is accomplished by applying a diagonal transformation to the unscaled structure. If we represent scaled quantities with a tilde, the scaled (transformed) matrices $\tilde{\Psi}_q, \ldots, \tilde{\Psi}_1$ can be related to the original unscaled matrices $\Psi_q, \ldots, \Psi_1$ as follows:

$$\tilde{\Psi}_i = S_i \Psi_i (S_{i-1})^{-1} \qquad \text{for } i = q, \ldots, 1, \tag{5.14}$$

where

$$S_0 = \begin{bmatrix} S_q & 0 \\ 0 & 1 \end{bmatrix}$$

and all S_i are diagonal. [The unit entry in S_0 simply reflects the fact that $y(k)$ does not change due to the scaling.] The entries in the S matrices are then derived from the K matrices computed in (5.12) and (5.13):

$$[S_i]_{jj} = ([K_i]_{jj})^{-1/2} \qquad \text{for } i = 1, \ldots, q \quad \text{and all } j. \tag{5.15}$$

The modified state space for the scaled structure would be

$$\begin{bmatrix} \tilde{v}(k+1) \\ \tilde{u}(k+1) \end{bmatrix} = \tilde{\Psi}_q \cdots \tilde{\Psi}_1 \begin{bmatrix} \tilde{v}(k) \\ \tilde{u}(k) \\ y(k) \end{bmatrix}. \tag{5.16}$$

Now if we computed the $\tilde{K}_i$ matrices for this scaled structure, they would all have unity diagonal elements, as desired. One more point concerning (5.16) must be mentioned. Our unscaled structure had input y and output u. However, the output node was scaled, and thus u does not appear in (5.16). To produce u, the D/A converter that follows the scaled output node must include an extra multiplicative factor ρ equal to the reciprocal of the $(n+1, n+1)$th entry of S_q.

5.2 Digital Feedback Compensator Scaling

In this section we shall discuss the implications of LQG set-point configurations for the issue of compensator scaling and then adapt the l_2 stochastic scaling method described in the previous section for filters to the digital feedback compensator.

The scaling issue for digital compensators differs in certain respects from the filtering applications described. The first of these involves the type of scaling appropriate to LQG systems. It is not uncommon for the LQG configurations described in chapter 2 to have *set points*, in other words, reference inputs for the regulator portion of the design. These nonzero set-point regulators [1] will have the same parameter values as the regulators described in chapter 2, independent of the set point, but the resulting DC compensator input will affect the scaling. As stated before, conservative scaling is required whenever we allow the presence of DC inputs. Specifically, l_2 scaling is not possible. From the stochastic viewpoint, the use of l_2 scaling would be meaningless; the probability of overflow at the various compensator nodes would actually vary as a function of the input reference level.

Figure 5.2 presents the set-point LQG system described in Kwakernaak

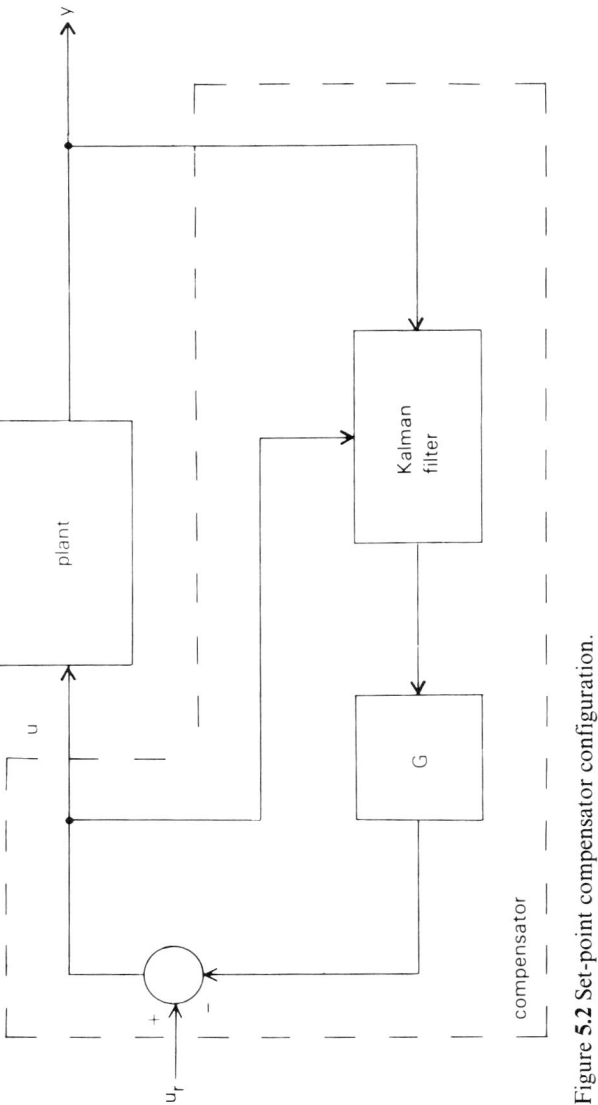

Figure **5.2** Set-point compensator configuration.

and Sivan [1], where u_r is the reference input. (Neglect driving and measurement noises.) If we wish to drive the output y to y_r, then u_r must be set to $H_c^{-1}(1)y_r$, where $H_c(z)$ is the closed-loop transfer function from u_r to y:

$$H_c(z) = L(zI - \Phi + \Gamma G)^{-1}\Gamma. \tag{5.17}$$

Unfortunately, the compensator in figure 5.2 essentially has a DC input since the steady-state value of y is nonzero. Thus l_2 scaling is not possible. Furthermore, the digital compensator in figure 5.2 has *two* inputs, which will require more complicated analyses than the single-input case.

Fortunately, there is one other (equivalent) approach to describing the system of figure 5.2 and the equations of chapter 2. Define ξ, η, and γ to be the *deviations* of the states, input, and output from the steady-state values x_0, u_0, and y_0. Thus $\xi = x - x_0$, $\eta = u - u_0$, and $\gamma = y - y_0$. As in [1], the following relation must hold:

$$\begin{aligned} x_0 &= \Phi x_0 + \Gamma u_0, \\ y_0 &= L x_0. \end{aligned} \tag{5.18}$$

Now follow through the LQG design equations of chapter 2 for the (deviations of the) states ξ, input η, and output γ. With the *actual* state, input, and output variables being represented by x, u, and y, we can then produce figure 5.3. Thus it is possible to use an alternative LQG set-point configuration in which the compensator input has no DC component, thereby allowing us to apply stochastic (l_2) scaling. The disadvantage to this alternative configuration is the necessity of having *two* reference inputs that must maintain the precise relation (5.18), typically in the presence of plant parameter uncertainty.

This disadvantage will vanish whenever the plant has a series integration (at least one pole at the origin $s = 0$, or $z = +1$ in the discrete domain), which is a very common occurrence in control systems. In fact, frequently an integrator is *added* to an actuator (part of the plant) to provide de-sensitivity to constant disturbances. To see the effect of an integrator pole on the configuration of figure 5.3, let us write u_0 as $(L(I - \Phi)^{-1}\Gamma)^{-1}y_0$ [from (5.18)]. However, since the gain $L(I - \Phi)^{-1}\Gamma$ blows up if there are any open-loop integrator poles in the plant (poles at $z = 1$), u_0 is forced to zero. In other words, if the plant has any series integration, the LQG configuration of figure 5.3 need have only one reference input, $y_0 = y_r$, and not two. Note that the configuration of figure 5.2 does not

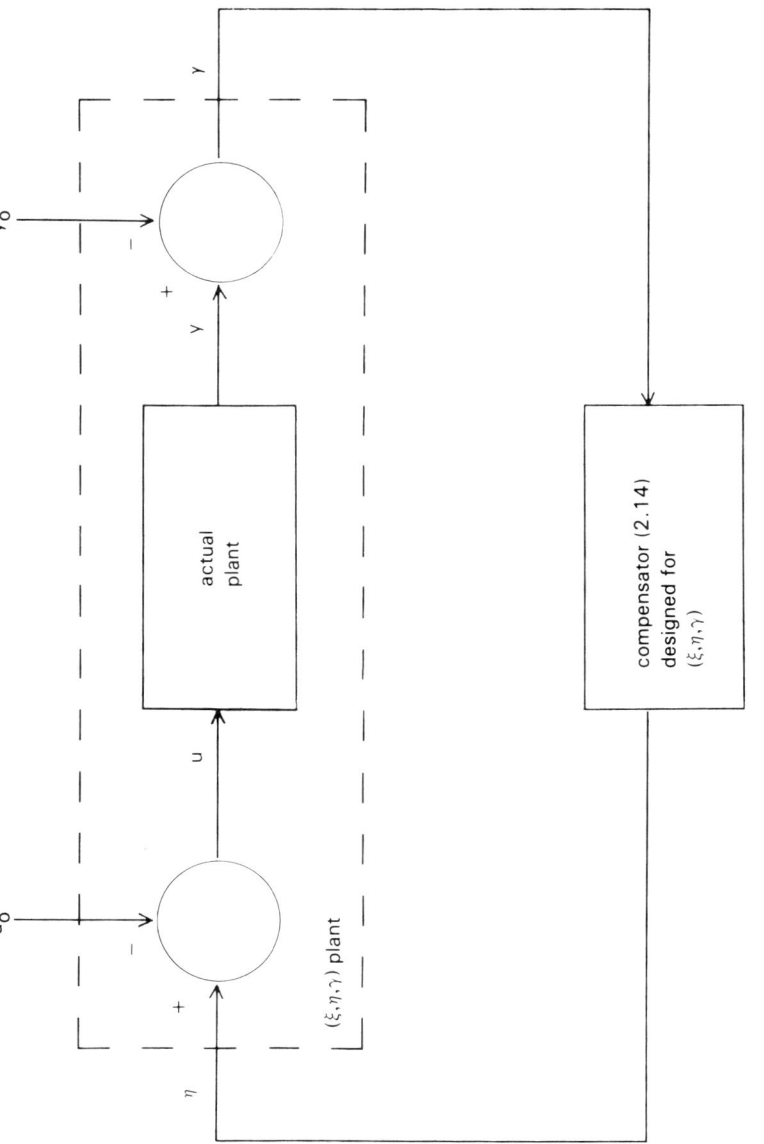

Figure **5.3** Alternative LQG set-point configuration.

change when the plant has integrator poles; both compensator inputs will still have DC components, and the system as a whole still requires the reference input u_r. From this point on, the configuration of figure 5.3 is assumed so that we may apply l_2 scaling and consider the compensator to be single-input. With this configuration we can now ignore the set point, since it will not affect our analyses. Thus we shall refer to the compensator input as y (not γ) and the output as u (not η).

The second difference between filter and compensator scaling arises when we try to apply l_2 scaling as described in (5.9)–(5.16) to a compensator. This procedure would treat the compensator as a separate entity (functionally a filter), ignoring the LQG plant and feedback path. Yet a compensator operating open-loop need not even be stable. (Consider a plant with a pole in the right half-plane at a higher frequency than a zero in the right half-plane among its singularities; in order to stabilize the plant, the compensator *must* have a pole in the right half-plane. The stochastic scaling method requires the variances of the signal variables at the compensator state nodes so that the matrices K_i and S_i can be computed. Clearly these variances will depend on the overall closed-loop system. From another point of view, the stochastic scaling described for filters is no longer valid because the compensator input y is no longer white. Thus we will have to *adapt* the filter scaling procedure so that it applies to digital feedback compensator scaling.

We have developed the following scaling procedure to account for the LQG feedback system in which the compensator is embedded. The steady-state variances of the n plant states and $n + 1$ compensator states can be found by combining the state and compensator equations into a single augmented state space:

$$
\begin{bmatrix} x(k+1) \\ v(k+1) \\ u(k+1) \end{bmatrix} = A \begin{bmatrix} x(k) \\ v(k) \\ u(k) \end{bmatrix} + \begin{bmatrix} w_1(k) \\ \Psi_{12} w_2(k) \end{bmatrix},
\tag{5.19}
$$

where

$$
A = \left[\begin{array}{c|cc} \Phi & 0_n & \Gamma \\ \hline \Psi_{12} L & & \Psi_{11} \end{array} \right]
$$

and 0_n represents an all-zero $n \times n$ matrix and Ψ_{11}, Ψ_{12} represent the unscaled compensator as partitioned in (5.9). With this state space, let us now follow the general scaling procedure outlined in section 5.1. The overall $(2n + 1) \times (2n + 1)$ state covariance matrix Z can be computed by solving the following discrete-time Lyapunov equation [16]:

$$Z = AZA' + C, \tag{5.20}$$

where

$$C = \begin{bmatrix} \Theta_1 & 0 \\ 0 & \{\Psi_{12}\Theta_2\Psi'_{12}\} \end{bmatrix}.$$

We can now partition Z to separate the plant and compensator co-variances:

$$Z = \begin{bmatrix} Z_{11} & Z_{12} \\ Z'_{12} & Z_{22} \end{bmatrix}, \tag{5.21}$$

where Z_{11} is $n \times n$. As defined in section 5.1, K_q will equal the compensator state covariance matrix Z_{22} divided by $\overline{yy} = \sigma_y^2$. Since $y = Lx$, this can be written

$$K_q = \frac{Z_{22}}{LZ_{11}L'}. \tag{5.22}$$

To compute K_i as in (5.13), the compensator states and input y would have to be uncorrelated. However, feedback introduces correlation:

$$E\left\{ \begin{bmatrix} v \\ u \\ y \end{bmatrix} [v' \; u \; y] \right\} = \begin{bmatrix} Z_{22} & Z'_{12}L' \\ LZ_{12} & LZ_{11}L' \end{bmatrix}. \tag{5.23}$$

If we take this into account, and again normalize by σ_y^2, we get the following equation for $i = 1, \ldots, q - 1$, which can be used in place of (5.14):

$$K_i = \Psi_i\Psi_{i-1} \cdots \Psi_1 \begin{bmatrix} K_q & \dfrac{Z'_{12}L'}{LZ_{11}L'} \\ \dfrac{LZ_{12}}{LZ_{11}L'} & 1 \end{bmatrix} \Psi'_1 \cdots \Psi'_{i-1}\Psi'_i. \tag{5.24}$$

The scaling matrices S_i now follow directly from (5.15). All of the compensator structures treated in the remainder of this work have undergone the optimal l_2 scaling just described.

The last controller scaling question that arises concerns the scale factors that are implicit in the A/D and D/A converters shown in figure 3.1.[If such scale adjustments are not available with specific converters, additional (analog) amplifiers or attenuators would probably have to be added.] Once a compensator is scaled via (5.19)–(5.24), the probability of overflow within the compensator equals the probability of overflow at the A/D input (ignoring set-point transients). By setting the A/D scale factor k_{ad} and inversely adjusting the D/A scale factor k_{da} (which already includes a term ρ due to the scaling of the compensator output node), we can set this overflow probability to an acceptable level without changing the ideal system response. It is also important to note that the compensator scaling procedure will be unaffected by this A/D scale factor—the scaling multipliers will remain invariant. Thus our compensator scaling procedure remains valid regardless of the A/D scaling. Besides the desired probability of overflow, the dynamic range of the input and output transients in the system (caused by changing the set point for example) will also affect the actual A/D scaling choice. However we choose k_{ad}, the effect of quantization noise on the performance index or on the output noise variance will increase (degrade) as k_{ad}^{-2}. For the analyses presented in the following chapters, k_{ad} will be taken to be one for simplicity; nonunity values can be accounted for by dividing the computed increase in J by k_{ad}^2 or by increasing the required wordlengths by $-\log_2(k_{ad})$ bits.

5.3 Quantizer Characteristics and Models

In order to analyze the effects of quantization in some tractable and systematic fashion, it is necessary to model the nonlinear operation of quantization. This section will present the roundoff and sign-magnitude truncation quantizer input-output characteristics and the models commonly used for them. A discussion of model validity then follows. Assume throughtout with no loss of generality that the fixed-point words representing signal variables have n_f bits, all to the right of the binary point, and that Δ is defined to be the quantization step size $\Delta = 2^{-n_f}$.

Figure 5.4 shows the input-output characteristic of the roundoff

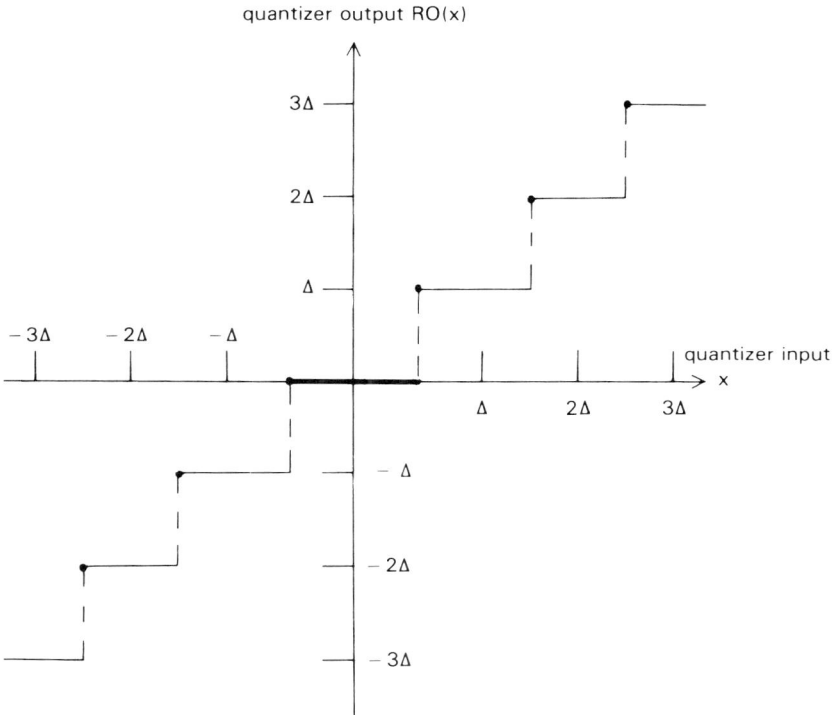

Figure **5.4** Nonlinear roundoff characteristic.

quantizer. Let $RO(x)$ be the rounded value of a quantity x. The error associated with such a quantizer, $e = x - RO(x)$, statisfies (5.25):

$$-\frac{\Delta}{2} < e \leqslant \frac{\Delta}{2}. \tag{5.25}$$

The model commonly used to represent this roundoff quantization operation is the additive white noise model [59]. In this case, roundoff is modeled *linearly* as a zero-mean uncorrelated noise added to the ideal (infinite-precision) signal value. The noise e is assumed to have a uniform density, as shown in figure 5.5, and to be uncorrelated with the quantizer input signal. The validity of this model is an important consideration, since its use simplifies quantization noise analysis a great deal. For a continuous-time quantizer input signal, the usually applied rule of thumb

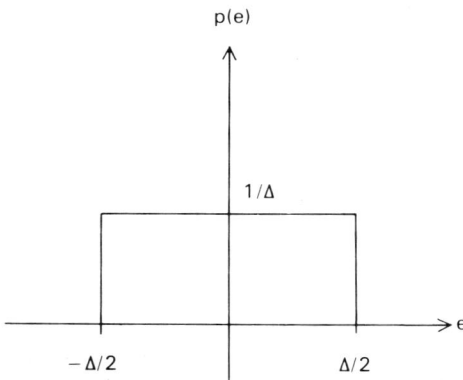

Figure **5.5** White noise error model density.

states that the noise model is valid if the input to the quantizer crosses
"many" quantization levels between sample times [29]; that is, the input
magnitude must fluctuate over a range $\gg \Delta$ in each T-second period.

A detailed analysis of the validity of the additive noise roundoff
model has been carried out by Sripad and Snyder [60] and Sripad [13].
These authors have established necessary and sufficient conditions on
the quantizer input such that the model is *exact*. Let $\phi_x(s)$ be the characteristic function of the quantizer input x [the Laplace transform of the
probability density $p(x)$]. Then

1. The density $p(e)$ matches that of figure 5.5 if and only if $\phi_x(2\pi i/\Delta) = 0$
for $i \neq 0$ and i an integer.

2. The noise samples $e(k)$ and $e(k + 1)$ are uncorrelated if and only if
the joint characteristic function between the two inputs $x(k)$ and $x(k + 1)$
satisfies $\phi_{x(k),x(k+1)} (2\pi i/\Delta, 2\pi j/\Delta) = 0$ for all $j, i \neq 0$.

3. The quantities $e(k)$ and $x(k)$ are uncorrelated if $\phi_x(2\pi i/\Delta) = 0$ and
$d\phi_x(\omega)/d\omega = 0$ for $\omega = 2\pi i/\Delta$ and all $i \neq 0$.

Unfortunately these conditions are difficult to verify since the probability
density function of *every* quantizer input must be known. Moreover, if a
quantizer input contains any Gaussian noise (a reasonable assumption
for control problems, especially at the A/D input quantizer), then we
know that the characteristic function will not exactly satisfy the conditions
required in (1) and (3).

This validity problem is not as serious as it seems. Sripad [13] has investigated the properties of the quantization error given a Gaussian input of variance σ^2. From these results it is evident that the error $e(k)$ has an approximately uniform distribution for $\sigma \gtrsim 0.7\Delta$, a condition that is not particularly restrictive.

In considering multiple quantizers (which is the usual case), the question of the interaction of the quantization errors arises. The analysis just given actually applies to a single quantizer only. When the model is used for all the quantizers within a complex (recursive) structure, we further assume that all such noise sources are independent. The question of the validity of this assumption is even more complex. However, it *can* be said that as a general technique, the additive noise model has proved itself quite useful for the analysis of roundoff noise effects in digital filters. Furthermore, any analysis techniques aimed at selecting word-lengths based on the effects of quantization noise need not be exact anyway—the internal and A/D wordlengths can only be selected in units of whole bits. When the roundoff noise model does break down, it tends to do so in a major way: limit cycles occur. These oscillations are usually quite evident when they are present (see chapter 7). For the following analyses, however, assume that the uncorrelated additive white noise model applies.

The second type of quantization that we shall discuss is sign-magnitude truncation. In sign-magnitude truncation, the extra bits of precision in the quantizer input are simply dropped. The advantages to this type of quantization are its simplicity—no extra hardware is required to implement sign-magnitude truncation, unlike the roundoff case—and its limit cycle behavior—fewer occur than in the roundoff case. Figure 5.6 shows the input-output characteristic of this quantizer. The quantization errors are now bounded as follows;

$$0 \leqslant e < \Delta \qquad \text{for } x \geqslant 0,$$
$$-\Delta < e \leqslant 0 \qquad \text{for } x \leqslant 0. \tag{5.26}$$

For this type of quantization, the modeling problem is more difficult. From (5.26) we can see that a definite correlation exists between the error and the input values with sign-magnitude truncation; $e(k)$ is a function of the sign of the quantizer input $x(k)$. Such noise in a digital structure is termed *state-dependent* noise. Although Sripad [13] does present an additive white noise model for this quantization operation, the conditions for which the model is valid are too restrictive for general

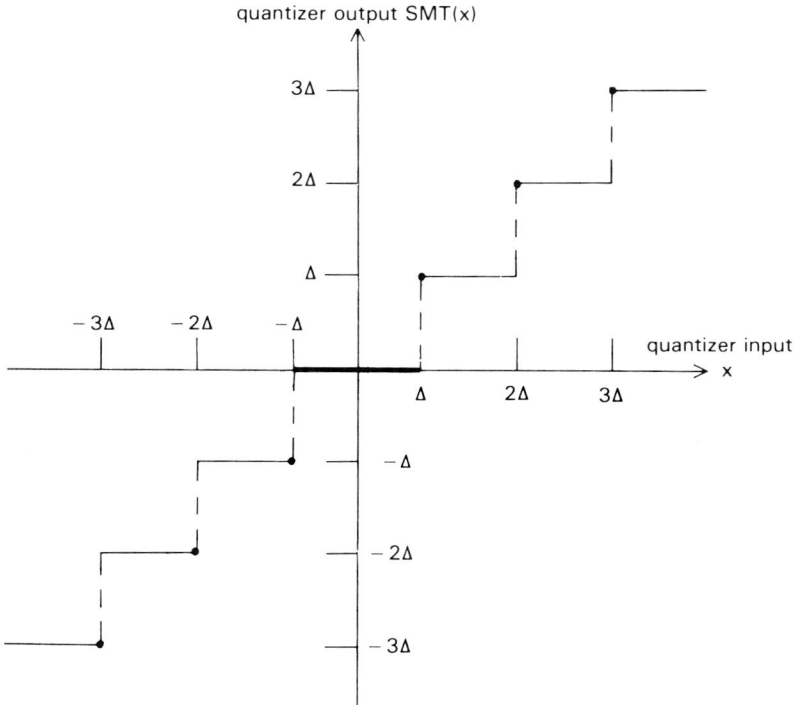

Figure **5.6** Nonlinear sign-magnitude truncation characteristic.

application. The additive white noise model is not even approximately valid, as is the roundoff model. Claasen, Mecklenbräuker, and Peek [61] have proposed a quasi-linear model for sign-magnitude truncation:

$$SMT(x) \approx x - \frac{\Delta}{\sqrt{2\pi}\sigma}x + e, \tag{5.27}$$

where e is an uncorrelated white noise of variance $(1/3 - 1/(2\pi))\Delta^2$ and the quantizer input x is assumed to be a Gaussian process. The dependence on σ (the variance of x) accounts for the quasi-linearity and also the complexity in using this model for analysis, since the variance of each quantizer input must be computed. An efficient technique for evaluating these variances is given in [61].

Empirically, the noise variance at the output of a digital filter using sign-magnitude truncation would typically be about 5 to 10 times that of the same filter using roundoff quantization [62]. Thus one should have an extra two bits per signal word when using sign-magnitude truncation in order to produce the same (or better) noise performance as would result from using roundoff quantization. Beyond this qualitative statement, we shall not consider the specific analysis of sign-magnitude quantization noise effects for control compensators.

5.4 Roundoff Noise Analysis

In this section we shall examine several methods for evaluating the effects of quantization noise in digital filters and compensators. The focus, as mentioned will be on *roundoff* quantization, using the additive white noise model. For filtering applications, we are typically concerned with the statistical effects of quantization on the filter output. Although Jackson [63] examines various norms of the output noise spectrum, the noise variance (the L_2 norm squared) is usually taken to be the metric.

There are two basic methods for computing the output variance resulting from quantization noise effects: one in the frequency domain and one in the time domain. The frequency-domain analysis method is an application of residue theory [64]. Given the ith noise source of variance $\Delta^2/12$ and the (scaled structure) transfer function $H_i(z)$ from the noise source to the output node, the output variance σ_i^2 due to this noise source can be written

$$\sigma_i^2 = \frac{\Delta^2}{12(2\pi j)} \oint H_i(z) H_i(z^{-1}) z^{-1} dz, \tag{5.28}$$

where j represents the square root of -1. The contour integral (5.28) can be evaluated by first factoring $H_i(z) H_i(z^{-1}) z^{-1}$ to determine its pole locations. If n_p of these poles z_l lie *inside* the unit circle, then

$$\sigma_i^2 = \frac{\Delta^2}{12} \sum_{l=1}^{n_p} (\text{Residue } \{H_i(z) H_i(z^{-1}) z^{-1}\} \text{ at } z_l). \tag{5.29}$$

Since every noise source is assumed to be uncorrelated with every other, the total output variance will simply be the sum of all the σ_i^2.

If we apply this residue method to the A/D quantization noise source, we see that σ_{ad}^2 will depend on the overall filter transfer function. Since each candidate structure by definition must have this same transfer function, given infinite precision coefficients, the effect of the A/D roundoff noise on filter output variance is dependent only on k_{ad} and the A/D wordlength. For a compensator the effect of A/D roundoff noise on J is similarly structure independent, given infinite-precision coefficients.

The time-domain approach to analyzing roundoff effects is presented by Hwang [65] for one-level state-space structures and Chan [17] for the general multilevel case. In the context of the modified state-space representation as presented in (3.17), the derivation proceeds as follows. Assume that the structure has already been scaled so that the factor ρ described after equation (5.16) must be included to produce $u(k)$ from the scaled $\tilde{u}(k)$. For a filter of input y, scaled output $\tilde{u}$, and scaled states $\tilde{v}$ [see (5.16)], the effect of roundoff noise on the filter states can be described by

$$\begin{bmatrix} \tilde{v}(k+1) \\ \tilde{u}(k+1) \end{bmatrix} = \tilde{\Psi}_{11} \begin{bmatrix} \tilde{v}(k) \\ \tilde{u}(k) \end{bmatrix} + \sum_{i=2}^{q} \tilde{\Psi}_q \cdots \tilde{\Psi}_i \varepsilon_{i-1}(k) + \varepsilon_q(k) + \tilde{\Psi}_{12} \varepsilon_{ad}(k),$$

(5.30)

where $\varepsilon_i(k)$ represents the noise sources due to the product quantizations associated with the precedence level matrix $\tilde{\Psi}_i$, the matrices $\tilde{\Psi}_{11}$ and $\tilde{\Psi}_{12}$ are defined analogously to those in (5.9) but for the scaled system, and $\varepsilon_{ad}(k)$ represents the A/D noise source. Recall that all such error sources are assumed to be uncorrelated. Thus the roundoff noise covariances can be written

$$\overline{\varepsilon_i(k)\varepsilon_i'(k)} = \frac{\Delta_r^2}{12} \Lambda_i,$$

(5.31)

$$\overline{\varepsilon_{ad}^2} = \frac{\Delta_{ad}^2}{12},$$

where Δ_r is the internal quantization step size of the structure, Δ_{ad} is the A/D quantization step size, and Λ_i is an $(n+1) \times (n+1)$ diagonal matrix whose (j,j)th entry equals the number of noninteger coefficients in the jth row of $\tilde{\Psi}_i$, that is, the number of roundoff error sources associated with the jth component of $r_i(k)$. This expression assumes that roundoff occurs after every nontrivial product. If double-precision adders are used as described in the introduction to this chapter, then we simply replace all the nonzero entries of Λ_i in (5.31) with ones.

To use (5.30) and (5.31) in computing the output variance, we can take either of the approaches used in section 5.1 for computing variances; that is, either the infinite series of (5.11) or the Lyapunov equation of (5.12) can be used. For the infinite-series approach, we would have to approximate the series by computing only a finite number of terms. The closer to the unit circle any of the poles of the system (5.31) are, the more terms will be required for an acceptable approximation [65]. Consequently, we shall use the Lyapunov equation method.

The steady-state (scaled) state covariance matrix $\tilde{V}$ can then be computed by solving the following Lyapunov equation [derived from (5.12), (5.30), and (5.31)]:

$$\tilde{V} = \tilde{\Psi}_{11} \tilde{V} \tilde{\Psi}'_{11} + \tilde{\Omega}, \tag{5.32}$$

where

$$\tilde{\Omega} = \tilde{\Psi}_{12} \frac{\Delta_{ad}^2}{12} \tilde{\Psi}'_{12}$$

$$+ \frac{\Delta_r^2}{12} \{ \Lambda_q + \tilde{\Psi}_q \Lambda_{q-1} \tilde{\Psi}'_q + \tilde{\Psi}_q \tilde{\Psi}_{q-1} \Lambda_{q-2} \tilde{\Psi}'_{q-1} \tilde{\Psi}'_q$$

$$+ \cdots + \tilde{\Psi}_q \cdots \tilde{\Psi}_2 \Lambda_1 \tilde{\Psi}'_2 \cdots \tilde{\Psi}'_q \}.$$

The output variance of $\tilde{u}$ will simply be equal to the lower right-hand-corner entry of $\tilde{V}$. As previously mentioned, these equations for roundoff analysis are solved using infinite-precision coefficients for simplicity. The insertion of the actual finite-wordlength coefficients would only change the results in a minor way (see section 6.6). In that case, there would be also a slight dependence on structure for the A/D noise contribution. The use of infinite-precision coefficients is especially justified when one recalls that the selection of an internal or A/D wordlength can only be made in terms of whole bits.

Now let us adapt this approach for the digital feedback compensator. Again, it is necessary to consider the behavior of the closed-loop system, as done by Knowles and Edwards [7] and Curry [8] for sampled-data systems and by Sripad [13]. Curry [8] has considered the second moment of the system output error due to rounding for a specific sampled-data control system with a direct form II compensator structure. Knowles and Edwards [7] also used the additive white noise model for generating a bound on the quantization noise effects of direct form II, cascade, and parallel compensator structures. Sripad [13] considered the increase in the performance index J due to roundoff, using the additive white noise model, but did not consider either the scaling issue or an accurate and general notion of a compensator structure. The results presented here will be more general since any type of compensator structure can be considered, and they will of course be adapted from the digital filtering approach just described.

Consider the scaled, augmented plant-compensator system, including roundoff noise sources, but not plant or measurement noises:

$$
\begin{bmatrix} x(k+1) \\ \tilde{v}(k+1) \\ \tilde{u}(k+1) \end{bmatrix} = \tilde{A} \begin{bmatrix} x(k) \\ \tilde{v}(k) \\ \tilde{u}(k) \end{bmatrix} + \begin{bmatrix} 0 \\ \varepsilon_q(k) + \sum_{i=2}^{q} \tilde{\Psi}_q \cdots \tilde{\Psi}_i \varepsilon_{i-1}(k) + \tilde{\Psi}_{12} \varepsilon_{\text{ad}}(k) \end{bmatrix},
$$
(5.33)

where

$$
\tilde{A} = \left[\begin{array}{c|c} \Phi & 0_n \;\vdots\; \Gamma\rho \\ \hline \tilde{\Psi}_{12} L & \tilde{\Psi}_{11} \end{array} \right].
$$

By superposition, we can consider only the roundoff noise sources, and thus compute only the *increase* in J due to these sources. We already know that the plant and measurement noise produce the ideal design value of J. By an approach similar to (5.32), the (scaled) state covariance matrix $\tilde{Z}$ (due only to roundoff noise) will be the solution to the following Lyapunov equation:

$$
\tilde{Z} = \tilde{A} \tilde{Z} \tilde{A}' + \begin{bmatrix} 0 & 0 \\ 0 & \tilde{\Omega} \end{bmatrix}.
$$
(5.34)

The covariance matrix $\tilde{Z}$ can then be related to the performance index J by using the *trace* form [27] of J, which is equivalent to (2.6):

$$
\begin{aligned}
J &= \text{trace}(Q\overline{xx'}) + 2\,\text{trace}(M\overline{ux'}) + \text{trace}(R\overline{uu}) \\
&= \text{trace } \Upsilon Z \\
&= \text{trace } \tilde{\Upsilon} \tilde{Z},
\end{aligned}
$$
(5.35)

where

$$
\tilde{\Upsilon} = \begin{bmatrix} Q & 0_n & M\rho \\ 0_n & 0_n & 0 \\ M'\rho & 0 & R\rho^2 \end{bmatrix}
$$

and

$$\Upsilon = \begin{bmatrix} Q & 0_n & M \\ 0_n & 0_n & 0 \\ M' & 0 & R \end{bmatrix}$$

By solving (5.34) and evaluating (5.35) for the scaled system covariance matrix $\tilde{Z}$, we can compute the increase dJ due to *roundoff noise alone*. Again, the infinite-precision coefficient values of the structure are used.

The analysis procedure described above extends easily to multiple-input multiple-output structures, but as described in chapter 9, the scaling issue is more complex.

5.5 Minimum Roundoff Noise Structures

Now that an analytic technique for treating roundoff noise effects has been presented, both for digital filters and for digital compensators, minimum roundoff noise structures can be described (see chapter 3). First, we shall describe the one-level minimum roundoff noise *filter* structure derived by Mullis and Roberts [18, 39, 40] and Hwang [41], and then we shall adapt the technique to produce a one-level minimum roundoff noise compensator structure.

In the context of the modified state-space representation, assume that a one-level n-pole *filter* structure has been l_2 scaled using (5.9)–(5.16) and that the roundoff noise could be evaluated with (5.32). Let us also neglect the A/D contribution in (5.32), so that we deal only with structure-dependent terms. For one level, (5.32) can be rewritten to include scaling as follows:

$$\tilde{V} = S_1 \Psi_{11} S_1^{-1} \tilde{V} S_1^{-1} \Psi'_{11} S_1 + \frac{\Delta_r^2}{12} \Lambda_1. \tag{5.36}$$

Recall that Λ_1 is an $(n + 1) \times (n + 1)$ diagonal matrix whose jth diagonal entry equals the number of roundoff error sources represented in the jth row of Ψ_1 and that the scaling matrix S_1 is diagonal. The variance at the output node u due to product quantization can be expressed with the following trace:

$$\sigma_0^2 = \rho^2 \, \text{trace} \, \Pi \tilde{V}, \tag{5.37}$$

where Π is an $(n + 1) \times (n + 1)$ matrix with a one in the $(n + 1, n + 1)$th location and zeros elsewhere.

By substituting V for $S_1^{-1} \tilde{V} S_1^{-1}$, we can rewrite (5.36) and (5.37):

$$V = \Psi_{11} V \Psi_{11}' + \frac{\Delta_r^2}{12} (S_1)^{-1} \Lambda_1 (S_1)^{-1}$$

$$= \Psi_{11} V \Psi_{11}' + \frac{\Delta_r^2}{12} \Lambda_1 S_1^{-2}, \tag{5.38}$$

$$\sigma_0^2 = \rho^2 \operatorname{trace}(\Pi S_1 V S_1)$$
$$= \operatorname{trace} \rho^2 S_1 \Pi S_1 V$$
$$= \operatorname{trace}(\Pi V). \tag{5.39}$$

Using the theory of adjoint operators [1], computing σ_0^2 via (5.38) and (5.39) is exactly equivalent to solving the following adjoint Lyapunov equation and evaluating the trace of (5.41) (see appendix B):

$$W_1 = \Psi_{11}' W_1 \Psi_{11} + \Pi, \tag{5.40}$$

$$\sigma_0^2 = \frac{\Delta_r^2}{12} \operatorname{trace}(\Lambda_1 S_1^{-2} W_1)$$

$$= \frac{\Delta_r^2}{12} \sum_{i=1}^{n+1} [\Lambda_1]_{ii} [K_1]_{ii} [W_1]_{ii}. \tag{5.41}$$

This alternative expression for roundoff noise will be important in the development of an iterative constrained optimization technique for minimizing roundoff noise effects, both for filter structures (see Chan [17]) and for compensator structures (see chapter 8).

Using the expression in (5.40), and the Lyapunov equation (5.12) for K_1, Mullis and Roberts [18] and Hwang [41] present a method for determining the similarity transformation that produces a structure with minimum σ_0^2. The matrix Λ_1 is assumed to be the identity I (for double-precision adders) or $(n + 1)I$ (for the case of single-precision adders and $n + 1$ rounded products per adder). Since the context of the modified state-space representation requires the output node to be a state, we cannot allow the output node to undergo a general similarity transformation. Thus we will confine the minimizing transformation to the n states v. Consequently, the $i = (n + 1)$th term in the summation expression for σ_0^2 in (5.41) will be fixed. In this case, we can ignore this term and deal only with K and W, the upper left-hand $n \times n$ portions of K_1 and W_1. Thus we must minimize the following sum:

$$\sum_{i=1}^{n} [K]_{ii}[W]_{ii}. \tag{5.42}$$

If P is an $n \times n$ (similarity) transformation matrix, then the product KW can be shown to transform to $P^{-1}KWP$. Thus the n eigenvalues of $(P^{-1}KWP)$ are invariant under transformation by P. These eigenvalues are called the second-order filter modes μ_i^2. Mullis and Roberts [39] prove the following inequality. If K and W are $n \times n$, symmetric positive-definite matrices, then

$$\frac{1}{n} \sum_{i=1}^{n} K_{ii} W_{ii} \geqslant \left[\frac{1}{n} \sum_{i=1}^{n} \mu_i \right]^2. \tag{5.43}$$

An (optimal) structural transformation exists [39] such that the transformed K_t and W_t $[K_t = P^{-1}K(P')^{-1}, W_t = P'WP]$ satisfy (5.43) with equality. Thus the minimum roundoff noise possible, using $(n+1)^2$ coefficients (in general) and quantization after every nontrivial multiplication, can be expressed as

$$(\sigma_0)_{\text{opt}}^2 = \frac{\Delta_r^2}{12} \left\{ (n+1)[K_1]_{n+1,n+1}[W_1]_{n+1,n+1} + \frac{n+1}{n} \left(\sum_{i=1}^{n} \mu_i \right)^2 \right\}, \tag{5.44}$$

assuming we know some initial K_1 and W_1 and can solve for the eigenvalues of KW.

If in fact we restrict ourselves to the *block optimal* parallel structure [39] with its (fewer) $4n + 1$ coefficients, then we are constraining the transformation P to be block diagonal and (5.43) cannot in general be satisfied with equality. However, (5.43) will be true for each *second-order section* $(n = 2)$. Thus the minimum block optimal product variance can be written

$$(\sigma_0^2)_{\text{bo}} = \frac{\Delta_r^2}{12} \left\{ (n+1)[K_1]_{n+1,n+1}[W_1]_{n+1,n+1} + \frac{3}{2}(\mu_1 + \mu_2)^2 \right.$$

$$\left. + \frac{3}{2}(\mu_3 + \mu_4)^2 + \cdots \right\}. \tag{5.45}$$

This equation in fact suggests a new result—a pairing algorithm for real poles. Once the modes of KW are determined, (5.45) will be minimized by pairing modes so that each pair of modes sums to approximately the same quantity as every other pair. In fact, (5.45) may be lower than (5.44) due to the reduced number of coefficients (noise sources).

This one-level minimum roundoff noise structure can be extended to the case of one-level compensators. Again, we may neglect the A/D noise contribution, which is invariant to structural transformation (assuming infinite-precision coefficients). Equation (5.34) can be rewritten in terms of its *unscaled* compensator parameters Ψ_{11} and Ψ_{12} as follows:

$$\tilde{Z} = TAT^{-1}\tilde{Z}T^{-1}A'T + \frac{\Delta_r^2}{12}\begin{bmatrix} 0 & 0 \\ 0 & \Lambda_1 \end{bmatrix}, \tag{5.46}$$

where

$$T = \begin{bmatrix} I_n & 0 \\ 0 & S_1 \end{bmatrix}$$

and

$$A = \begin{bmatrix} \Phi & 0_n & \Gamma \\ \hline \Psi_{12}L & \Psi_{11} \end{bmatrix}.$$

By recognizing that the unscaled covariance matrix Z just equals $T^{-1}\tilde{Z}T^{-1}$, we can write (5.46) in a manner similar to (5.38) to produce

$$Z = AZA' + \frac{\Delta_r^2}{12}\begin{bmatrix} 0 & 0 \\ 0 & \Lambda_1 S_1^{-2} \end{bmatrix}. \tag{5.47}$$

The expression for the increase in performance index due to roundoff noise for the scaled system can also be written in terms of the unscaled covariance matrix Z [see (5.35)]:

$$\begin{aligned}
dJ &= \text{trace}\{\tilde{\Upsilon}\tilde{Z}\} = \text{trace}\{\tilde{\Upsilon}T^{-1}ZT^{-1}\} \\
&= \text{trace}\{T^{-1}\tilde{\Upsilon}T^{-1}Z\} \\
&= \text{trace}\{\Upsilon Z\}.
\end{aligned} \tag{5.48}$$

Using an adjoint Lyapunov equation, as in (5.40) and (5.41), we can express (5.47) and (5.48) as follows:

$$W = A'WA + \Upsilon, \tag{5.49}$$

$$dJ = \frac{\Delta_r^2}{12}\text{trace}\left\{\begin{bmatrix} 0 & 0 \\ 0 & \{\Lambda_1 S_1^{-2}\} \end{bmatrix}W\right\}. \tag{5.50}$$

If we define W_1 to be the lower right-hand $(n+1) \times (n+1)$ portion of W, then

$$dJ = \frac{\Delta_r^2}{12} \text{trace}\{\Lambda_1 S_1^{-2} W_1\}. \tag{5.51}$$

This expression is identical to the expression in (5.41). From this point on, the derivation of a one-level minimum roundoff noise compensator structure is exactly the same as the Mullis and Roberts and Hwang procedure [see (5.40)–(5.45)].

Conceptually, the technique described could be extended to multiple levels. However, the iterative structure optimization procedure considered in chapter 8 is far more useful for minimizing roundoff noise.

5.6 The F8 Example and Compensator Roundoff Noise

This section will examine the roundoff noise and scaling associated with some of the structures discussed in chapter 3 for an actual sixth-order LQG system. This system is a simplified model of the longitudinal dynamics of the F8 fighter aircraft at flight condition 12 (an altitude of 20,000 feet and a speed of mach 8) [66]. Longitudinal control of the aircraft is restricted to the elevator alone and a single measurement y formed; these simplifications make the plant model single-input single-output, so that all our analysis techniques directly apply. The actual multiple-input multiple-output model could be considered with similar techniques, but certain additional issues arise as discussed in chapter 9.

First-order actuator dynamics are included in the plant model, and a series integrator is also added. Thus the configuration of figure 5.3 will only have one reference input. Appendix A presents the continuous-time plant model in detail. The sample rate (10 hertz) is selected to be well above the highest plant pole frequency (12 radians/second). Thus T equals 0.1 seconds. The resulting discrete-time model parameters are also shown in appendix A.

For this plant model, the design equations of chapter 2 were followed. The resulting matrices K and G are also given in appendix A. All calculations were done in double precision (16 digits, or 54 bits) so that the system parameters and the parameters K and G are effectively infinite-precision quantities. The resulting performance index J is 0.00176477. This number is then taken to be the *ideal* value of the performance index, and degradation is measured relative to it.

To five significant digits, the poles and zeros of the (ideal) compensator transfer function are given in table 5.1. Note that unlike higher-order

Table **5.1** F8 compensator poles and zeros

Pole frequencies		Zero frequencies	
z_{p1}	$= 0.29179$	z_{z1}	$= 0.30119$
z_{p2}	$= 0.58904$	z_{z2}	$= 0.96728$
z_{p3}	$= 0.99514$	z_{z3}	$= 0.99878$
z_{p4}	$= 0.99869$	$z_{z4}, z_{z5} = 0.88189 \pm j0.26766$	
$z_{p5}, z_{p6} = 0.73149 \pm j0.40220$			

digital filters, there are many *real* poles and zeros in this compensator. This fact complicates the pairing issue for parallel and cascade structures. Note also the presence of poles and zeros very near the unit circle at $z = +1$; these singularities can be critical in determining an acceptable structure.

Before discussing the different structures tested, the structure-independent A/D noise contribution will be considered. If we allow a 5% increase in J due to this single noise source, then the procedure outlined in (5.33)–(5.35) using only ε_{ad} results in a 4.98-bit A/D wordlength. (This number does not include the sign bit.) Typically for filtering applications, the A/D wordlength need not be as long as the structure's internal wordlength; the same result appears for this control and estimation application, as will be seen. It should be mentioned that the A/D scale factor for this system was selected to be 0.25, to reduce the number of overflows. This factor has already been taken into consideration (see section 5.2) in the results reported in this section.

Ten structures were evaluated in terms of their product roundoff noise effects on J: the direct form II, five parallel forms (including a block optimal structure), three cascade structures, and the simple structure of equation (3.25). The direct form II structure (a) has been described in figure 3.6 and equation (3.19), and has 13 coefficients, including a single scaling multiplier. The first parallel structure (b) is composed of five direct form II sections—one second-order section (for the complex pole pair) and four first-order sections. Each section requires its own scaler, so this structure has a total of 17 coefficients. The next two parallel structures use three second-order sections, and hence the issue of how we pair the four real poles into two sections must be addressed. (There are three different ways.) Parallel structure (c) pairs z_{p1} with z_{p4} and z_{p2} with z_{p3}, separating the two near-unit-circle poles, while structure (d) pairs these two poles (z_{p3} and z_{p4}) together (see appendix A). Each structure will require three scalers, for a total of 15 coefficients.

Structures (a)–(d) are all direct form II based and thus require two precedence levels. Parallel structure (e) is a one-level structure produced by computing $\Psi_\infty = \Psi_2 \Psi_1$, where Ψ_2 and Ψ_1 are from structure (c), and using the result as a one-level structure (see section 3.3). This structure will still be a parallel combination of second-order sections; each section will be a one-level version of a direct form II section. This structure will have 16 coefficients, one more than (d) or (e). Parallel structure (f) is a minimum roundoff noise block optimal structure as in equation (3.26) and uses the same pole pairing as parallel structures (c) and (e).

As mentioned in chapter 3, cascade structures involve the issues of pairing and ordering; in addition to the pairing issues encountered with the parallel structure, the zeros must be paired, and the sections must be ordered. Jackson [63] has described general section ordering and pairing criteria. Consider the ith second-order section:

$$H_i(z) = \frac{1 + a_{i1}z^{-1} + a_{i2}z^{-2}}{1 + b_{i1}z^{-1} + b_{i2}z^{-2}}. \tag{5.52}$$

Complex pole pairs and complex zero pairs that are *nearest* each other are placed in the same section (paired). *Nearness* means that we try to pair poles and zeros so as to minimize the peak magnitude (L_∞ norm) of $H_i(z)$ for all i. As for section ordering, when direct form II sections are used (with l_2 scaling), the noise variance of the filter output *tends* to be minimized by ordering the sections in terms of increasing κ_i, where

$$\kappa_i = \frac{\|H_i\|_\infty}{\|H_i\|_2}. \tag{5.53}$$

(This is not a precise result.) Different guidelines apply if the L_∞ norm of the output is the performance gauge, rather than the L_2 norm, or if the direct form II section is not used. Unfortunately, Jackson does not consider the pairing of *real* poles. Dehner [67] and Hwang [68] develop general suboptimal algorithms for selecting good pairing and ordering, but these methods tend to require significant computer time to figure out the ordering and pairing for higher-order filters, and they still do not address the pairing of real poles.

Our roundoff analysis will consider just two different cascade pairing-ordering combinations. Cascade structure (g) consists of an arbitrarily chosen arrangement of poles and zeros (see appendix A); section 1 contains the complex pole pair (z_{p5}, z_{p6}) and real zero z_{z1}; section 2 contains the near unit-magnitude real poles z_{p3} and z_{p4} and the complex zero pair (z_{z4}, z_{z5}); and section 3 contains the near unit-magnitude real zeros z_{z2} and z_{z3} with the real poles z_{p1} and z_{p2}. Cascade structure (h) splits the near unit-magnitude poles and zeros and puts the complex pole and zero pairs together in the same section (see appendix A). Both (g) and (h) require three scalers and a total of 15 coefficients and four precedence levels [see (3.21) and figure 3.7]. Cascade structure (i) has the same ordering and pairing as (g), but uses direct form I sections as described in (3.22) and figure 3.8. Hence it has *different* scaling than (g), different scaled coefficients, and fewer scalers.

Finally, the simple structure (j) of (3.25) is treated since this structure (or a one- or two-level version of it) has been often used, even though it requires (after scaling) an excessive 50 coefficients for the F8 system example.

Appendix A contains the actual modified state-space representations of all 10 of these (l_2-scaled) structures. The ideal values of these coefficients are presented in double precision.

Table 5.2 summarizes the product roundoff noise results (that is, the noise caused by the rounding of multiplier products, not A/D rounding) for these 10 structures, assuming optimal l_2 scaling and *not* accounting for the finite wordlengths of the coefficients themselves. The *levels* column

Table **5.2** Roundoff noise results

Structure	Levels	N	Wordlength		max/min l_2; scaled coefficient
			spa	dpa	
(a) Direct form II	2	13	19.65	18.25	60050/0.12
(b) Parallel direct form II	2	17	8.05	7.45	1.5/0.0046
(c) Parallel direct form II	2	15	10.18	9.39	10.5/0.073
(d) Parallel direct form II	2	15	14.74	13.94	15.7/0.0015
(e) Parallel, 1-level version of (c)	1	16	9.78	8.99	6.3/0.073
(f) Block optimal parallel	1	25	7.88	7.06	1.1/0.0029
(g) Cascade direct form II	4	15	15.69	14.68	1101/0.00052
(h) Cascade direct form II	4	15	10.51	9.47	35/0.073
(i) Cascade direct form I	3	14	15.52	14.36	320/0.012
(j) Simple	3	50	9.01	7.54	1.6/0.0000003

lists the number of precedence levels, and the N column lists the number of (nonunity) coefficients, including scalers, in the structure. The roundoff noise results are presented in terms of the number of signal (wordlength) bits that are required to hold the increase in J due to product roundoff noise to 5% of the ideal value. Again, these numbers do not include the required sign bit. Two wordlengths are presented for each structure. The (larger) left-hand column corresponds to the case of roundoff after every nontrivial multiplication and single-precision adders, while the right-hand column corresponds to the case of double-precision adders and quantization after addition. The last column of table 5.2 shows the maximum and minimum magnitude of the scaled coefficients and is important in determining the coefficient wordlength. The wider the range of values, the more fixed-point coefficient bits will probably be needed to achieve a given level of performance (see chapter 6).

From table 5.2 we can see that the different pole pairings associated with parallel structures (c) and (d) produced results that differed by 4.5 bits. Placing the near-unit magnitude poles in different sections was quite effective. Similarly, of the two cascades (g) and (h), the one with these same two poles in different sections required 5.2 fewer bits. Clearly the pairing-ordering issue is not a trivial question.

Structure (b), the combination of first- and second-order parallel sections, with its 17 coefficients outperformed every other structure except the block optimal. Even so, the extra 8 coefficients of the block optimal structure with second-order sections only gained 0.2 bits of performance over this structure. Thus, when evaluating different structures, it is important to know the block optimal result (for various pairings) so that we can judge whether a suboptimal structure like (b) is *sufficiently* effective. In this case it clearly is. If we are constrained to one level, then (e) is probably best, given that it requires 9 fewer coefficients than the optimal and has only 1.9 bits poorer performance. Actually, in this case one should check the performance of a one-level version of (b).

As expected from the literature on digital filters, the discrete form II has a very poor noise performance. It is interesting to note also that the simple structure with its many coefficients (and hence many noise sources) performed excellently. It is not clear whether this would be true for the simple structure in general.

The second wordlength column in table 5.2 shows the gain possible when using double-precision adders and fewer quantizers. Depending on the structure tested, a saving of 0.6 to 1.47 bits was realized. Whether this small saving is enough to justify the higher-precision adders will depend on the particular application.

5.7 Summary

The process of scaling a digital feedback compensator to satisfy dynamic range constraints requires the consideration of the overall closed-loop control system in which the compensator is embedded. Thus we had to adapt the methods developed for scaling digital filters to this problem. Furthermore, when applying the stochastic l_2 scaling approach to the set-point LQG system, an alternative configuration for the system had to be considered.

For the analysis of roundoff noise effects in compensators, we again had to adapt the techniques used in digital signal processing to consider the effects of the overall closed-loop system. The development of minimum roundoff noise structures for compensators required similar adaptations. When these methods were applied to a specific control system example, we were able to compare different types of structures in terms of their roundoff noise performance. The pairing and ordering issues involved with the parallel and cascade structures were shown to be even more complex for compensators, due to the numbers of real poles that are common in control system compensators. Furthermore, the default structure for LQG controllers, the *simple* form, was shown to be a poor choice of structure in general for the LQG compensator. Although it performed well, it required far too many coefficients.

6

Finite-Wordlength
Effects: Quantizing
the Coefficients

The implementation of a discrete-time system described by an ideal infinite-precision transfer function in finite-precision hardware involves several important issues. Chapter 5 has discussed the quantization noise problem, and chapter 7 will present the issue of limit cycle oscillations. This chapter will consider the problem of quantizing the infinite-precision coefficients of the structure so that they may be stored using a finite-length fixed-point binary representation. As with the roundoff noise question, coefficient quantization effects are also heavily structure dependent, and thus the analysis of such effects is important when selecting a good structure and its required coefficient wordlength.

Approximating the coefficients of a structure with a finite number of bits will cause a degradation in the system's performance as compared to the ideal. Assuming that a given quantitative performance measure is provided, we can measure the trade-off in the number of bits used to represent the coefficients versus the degradation in performance. Then, assuming that we have specified an acceptable amount of degradation, we can determine the minimum number of coefficient bits needed to meet this goal. Repeating this procedure for many different structures, we can determine the one that has the smallest such wordlength.

Whatever the structure, the fewest number of total coefficient bits will be required if we allow each coefficient to have a different wordlength. We certainly will not need fewer total bits after *adding* a constraint such as uniform wordlength. However, the resulting complication in the digital hardware due to nonuniform memory widths and restrictions on the hardware multipliers make this superior apportionment of coefficient bits very costly. For this reason a uniform fixed-point coefficient worldlength is typically assumed. This assumption will be carried throughout the analysis, assuming n_c fractional bits, a sign bit, and enough integer bits to represent the largest coefficient in the structure. We shall also assume that each structure has already been scaled, since the scaling operation can radically change the dynamic range of the coefficients, and hence the required wordlength. However, the A/D scale factor can be ignored, or set to 1, since this scale factor has no effect whatsoever on the required coefficient wordlength of a structure.

Another important issue concerns the way in which we quantize the ideal coefficients. The simplest and most common procedure is to *round* the coefficients to n_c fractional bits. Unfortunately, there is no guarantee that this is the best method in terms of some specific performance metric. In fact, the optimal set of coefficients with n_c fractional bits is usually *not* these rounded values. This fact has given rise to several optimization techniques [20, 69, 70] for determining the best set of quantized coefficients for a given structure and wordlength. Typically these techniques start *near* the rounded coefficient set (in discrete coefficient space) and search for minima. Unfortunately, these methods can be extremely time consuming, with the resulting coefficient set not necessarily that much better than that obtained by rounding. Consequently, we shall assume that finite-wordlength coefficients are produced by rounding the ideal values.

The remainder of this chapter is organized as follows. In section 6.1 we shall describe different methods for selecting structures that have small required coefficient wordlengths and different ways of evaluating the required coefficient wordlength once a structure is selected. In particular, we shall discuss a *statistical* approach to wordlength determination and structural comparison that has been adapted from the field of digital signal processing. The statistical wordlength estimate has a very important advantage over any other approach—it can be used as the objective function in an iterative structure optimization procedure (see chapter 8). Sections 6.2 and 6.3 describe the statistical method in detail for the LQG problem, while section 6.4 presents a direct wordlength evaluation procedure. Section 6.5, drawing on the F8 system presented in chapter 5, presents various coefficient wordlength results and conclusions. Finally, the joint analysis of coefficient wordlength effects and roundoff noise effects is addressed in section 6.6.

6.1 Methods of Analysis

In this section we shall discuss three methods of analyzing the effects of finite coefficient wordlengths. The first is a qualitative approach based on pole location density and can be used primarily for comparing different structures on a limited basis. The other two methods can be used both for structural comparison and for the determination of the wordlengths required to meet some accepted performance level. These are the direct, or brute-force, and statistical approaches.

The effect of a quantized coefficient on any performance measure is essentially a *sensitivity* question. From a frequency-domain viewpoint, having coefficients of finite wordlength implies that there are only a finite number of possible pole and zero locations in the z-plane. Thus one approach to the selection of a structure with minimal coefficient quantization effects could be accomplished by examining a graph, or grid, of these locations; the coefficient sensitivity in an area of high grid density would be small. Thus the structure that had the densest grid in the area of the desired poles and zeros would be chosen. Several structures have been described in terms of pole location grids; for example, the *coupled form* second-order section of Rader and Gold [71] has a uniform square grid over the entire z-plane, while the direct form II has a nonuniform grid, densest near $z = \pm j$. Avenhaus [36], Abu-El-Haija, Shenoi, and Peterson [72], and Agarwal and Burrus [73] have described second-order sections whose grids are densest near $z = +1$, thus making them excellent for implementing lowpass filters. Avenhaus has also presented other sections and their respective pole location grids. Such a general approach to filter structure selection at least has an intuitive appeal. Of course, there is no guarantee that a structure with high grid density for the desired pole and zero locations will necessarily be the *best* structure in terms of a specific measure of performance degradation due to coefficient quantization effects, such as the trace of the error covariance (for a Kalman filter), the performance index J (for LQG systems), or phase margin (for a classical control system). Furthermore, the method is ill suited to applications for which the pole and zero locations are not *all* in dense grid areas (typical of control applications).

Given any set of quantized coefficients, the most direct and accurate way to evaluate the effect of finite wordlength on performance would be to recompute *for the quantized coefficient values* the entire transfer function, performance index J, phase margin, or whatever quantitative measure is appropriate. In fact, this is the approach taken by Sripad [13] for analyzing the effects of finite-wordlength coefficients. While this method has the virtue of being accurate, it tells us only one point on the performance-wordlength trade-off curve. The performance measure would have to be reevaluated for each potential wordlength until the desired degradation level has been bracketed (bounded above and below) by wordlengths differing only by one bit. Then the larger of the two wordlengths would be the required coefficient wordlength for that structure. Such a brute-force approach could be quite time consuming, especially when we wish to compute the required number of bits for several candidate structures.

It would be quite convenient to have a procedure in which a single evaluation established the entire behavior of the performance-wordlength trade-off curve. The required wordlength could then be estimated easily from knowing the allowed degradation level. Also, since the actual word-length *must* be integral, some accuracy can be sacrificed to gain simplicity as long as the required wordlength is not *underestimated*. More important, if the coefficient wordlength estimate is continuous in nature, that is, not confined to an integral number of bits, then it is possible to apply an optimization technique [17] to synthesize better structures. In this procedure, which is described in chapter 8, continuous transformations are applied to an initial structure. These transformations are determined by a gradient search technique based on a continuous and differentiable scalar objective function of the coefficients of the structure. Certainly, if our required wordlength estimate is strictly integral, it will not be differentiable.

The concept of a *statistical* estimate of wordlength has both of the advantages mentioned. This approach originated in the study of digital filters with the work of Knowles and Olcayto [19]. Avenhaus [20] applied this idea to the digital filter power transfer function (as a performance measure), and later Crochiere [21, 34] used the concept with the filter transfer function magnitude and a wordlength-optimization procedure. In the remainder of this section we shall review the basic development of the statistical wordlength measure for digital filters [21].

Consider a general scalar measure of performance f that is a function of a set of coefficients and is continuous and differentiable. For example, the error in the transfer function magnitude at a specific frequency, the integrated squared error in the transfer function magnitude, and the performance index for a steady-state LQG problem would be acceptable measures. With a finite-precision implementation, the resulting f will depend on the N quantized coefficients $(c_1, c_2, \ldots, c_N)$ of the structure. The value of f associated with any particular structure with finite-precision coefficients will reflect a degradation in performance as compared to the ideal (based on infinite-precision coefficients) value f_∞. This degradation df can be expanded in a Taylor's series about the ideal value. Keeping only first-order terms,

$$df(c_1, c_2, \ldots, c_N) \approx \sum_{i=1}^{N} \left(\frac{\partial f}{\partial c_i} \bigg|_\infty dc_i \right), \tag{6.1}$$

where c_i is the ith coefficient to be quantized, dc_i is the error due to quantization, and $(\partial f / \partial c_i)|_\infty$ is the first partial derivative of f evaluated at the infinite-precision coefficient values. Note that integer coefficients or coefficients such as 0.5 and 0.25 are normally not affected by quantization and should not be included in the sum (6.1).

If Δ is the quantization step size 2^{-n_c}, the fraction represented by the least significant bit of the fixed-point coefficient word, then each dc_i must lie between $\pm \Delta/2$. Given the partial derivatives in (6.1), we could then upper bound the error df, producing a *very pessimistic* wordlength estimate:

$$df < \frac{\Delta}{2} \sum_{i=1}^{N} \left| \left(\frac{\partial f}{\partial c_i} \bigg|_\infty \right) \right|. \tag{6.2}$$

The basic idea behind statistical wordlength is to produce a less pessimistic estimate by treating an ensemble of structures. Over this ensemble, the coefficient errors dc_i (assuming rounding) can be thought of as uniformly distributed zero-mean *uncorrelated* random variables, each of variance $(\Delta^2)/12$ (recall figure 5.5). Using (6.1), we can now treat the error df as a random variable. With dc_i as described, the error df is then zero-mean with a variance

$$(\sigma_{df})^2 = \frac{\Delta^2}{12} \sum_{i=1}^{N} \left(\frac{\partial f}{\partial c_i} \bigg|_\infty \right)^2. \tag{6.3}$$

For large N, the central limit theorem can be applied to justify a Gaussian distribution for df. Thus with a given confidence level (probability), say 95%, one can determine the variance needed for the error df to remain within some prescribed bound. In other words 95 out of 100 of the structures in the ensemble will result in systems in which df remains within this bound.

From a table of the Gaussian distribution,

$$\Pr[\,|df| \leqslant 2\sigma_{df}\,] = 0.954. \tag{6.4}$$

If the quantity of interest f is constrained to lie within $\pm E_0$ (the degradation level) of the ideal f_∞, then (6.4) implies that σ_{df} equals $(E_0)/2$. This result can be combined with (6.3) to produce an estimate of the parameter Δ:

$$\Delta = \frac{\sqrt{3}E_0}{\left\{ \sum\limits_{i=1}^{N} \left(\left. \frac{\partial f}{\partial c_i} \right|_\infty \right)^2 \right\}^{1/2}}. \tag{6.5}$$

Given Δ, the statistical wordlength can be defined to be

$$SWL = l + \log_2 \frac{1}{\Delta}. \tag{6.6}$$

The first term in (6.6) represents the number of bits necessary to represent the integer portion of the coefficient word (bits to the left of the fixed binary point), and the second term gives the number of bits n_c necessary for the fractional portion of the coefficient word (bits to the right of the binary point). The sign bit is not included in this expression.

In the digital filter area, Crochiere [21, 33, 34] has presented a number of results comparing the statistical wordlength of structures using the transfer function magnitude as the performance measure f. Since this choice of f is frequency dependent, the resulting estimate is also frequency dependent. The final wordlength can be selected as the maximum of the estimates over the frequency range of interest. In the examples treated by Crochiere, the statistical wordlength estimate was 1 to 3 bits conservative as compared to the actual minimum number of bits necessary to just meet the transfer function error limit. In a related work by Chan and Rabiner [74], which considered a large number of finite impulse response filters and a similar statistical approach to coefficient wordlength, the resulting

95% confidence level estimates were also observed to be conservative. Crochiere [21, 34] and Chan [17] have also formulated different optimization procedures for designing filter structures with shorter required coefficient wordlengths. As we shall show in chapter 8, such optimizing synthesis techniques are a strong motivation for developing the statistical approach for compensators.

6.2 Statistical Wordlength and LQG Systems

As mentioned in chapters 1 and 2, it is natural to use the performance index J of (2.3) as the measure of performance f for a steady-state LQG system. It would also have been possible to select phase margin, or gain margin, or any other metric appropriate to a control system, although the analyses would probably be quite complex. Using the approach of the previous section, the change in J_∞ due to finite-precision coefficients would be estimated by

$$dJ(c_1, c_2, \ldots, c_N) \approx \sum_{i=1}^{N} \left(\frac{\partial J}{\partial c_i} \bigg|_\infty dc_i \right). \tag{6.7}$$

However, the optimal natural of the LQG control problem forces all the first-order sensitivities $(\partial J/\partial c_i)|_\infty$ to be zero. Therefore a higher-order approximation is necessary:

$$dJ \approx \frac{1}{2} \sum_{i=1}^{N} \sum_{j=1}^{N} \left(\frac{\partial^2 J}{\partial c_i \partial c_j} \bigg|_\infty dc_i \, dc_j \right). \tag{6.8}$$

The use of second-order terms (not used in digital filter analysis) is a unique aspect of our statistical wordlength formulation. However, the use of these terms would be implicit in any statistical estimate based on the error in a scalar performance measure that has been optimized with respect to the coefficients. For example, if a digital filter is designed by minimizing the integrated squared error between the desired and actual filter transfer function magnitude characteristic, then a statistical wordlength estimate *based on this performance measure* has to use second-order sensitivities—all first-order sensitivities are zero. The statistical wordlength estimate that we have developed has to be used in such a case, and therefore this formulation also has some potential applications in digital filter design. However, most digital filter examples do not fall into this category. Crochiere [21, 34] dealt with the case for which the overall filter statistical estimate was taken to be the maximum over a set of estimates made at specific frequencies (each based on the transfer function magnitude error at that frequency). The first-order sensitivities for each of those estimates are nonzero since filters are not in general designed to minimize error at just one frequency. Furthermore, quite often digital filters are not designed by minimizing a differentiable scalar criterion. Thus in most filtering applications, the statistical estimate developed by Crochiere is appropriate.

Proceeding from (6.8), we can derive expressions for the mean and variance of dJ. Again, assume that the errors dc_i and dc_j are uncorrelated for $i \neq j$. The mean of dJ will be nonzero, unlike the mean of df in section 6.1:

$$E(dJ) = \frac{1}{2} \sum_{i=1}^{N} \left(\frac{\partial^2 J}{\partial c_i^2} \bigg|_{\infty} \right) E[(dc_i)^2]. \tag{6.9}$$

Before deriving an expression for the variance of dJ, let us define the random variable ε to be the square of dc_i. Its mean and variance can be shown to be $E(\varepsilon) = \bar{\varepsilon} = (\Delta^2)/12$ and $E(\varepsilon^2) = \overline{\varepsilon^2} = (\Delta^4)/180$. The second moment and variance of dJ can now be written as follows:

$$E[(dJ)^2] = \frac{\overline{\varepsilon^2}}{4} \sum_{i=1}^{N} \left(\frac{\partial^2 J}{\partial c_i^2} \bigg|_{\infty} \right)^2 + \frac{(\bar{\varepsilon})^2}{4} \sum_{\substack{i=1 \\ i \neq j}}^{N} \sum_{j=1}^{N} \left(\frac{\partial^2 J}{\partial c_i^2} \bigg|_{\infty} \right) \left(\frac{\partial^2 J}{\partial c_j^2} \bigg|_{\infty} \right)$$

$$+ \frac{(\bar{\varepsilon})^2}{2} \sum_{\substack{i=1 \\ i \neq j}}^{N} \sum_{j=1}^{N} \left(\frac{\partial^2 J}{\partial c_i \partial c_j} \bigg|_{\infty} \right)^2, \tag{6.10}$$

$$(\sigma_{dJ})^2 = \frac{\sigma_\varepsilon^2}{4} \sum_{i=1}^{N} \left(\frac{\partial^2 J}{\partial c_i^2} \bigg|_{\infty} \right)^2 + (\bar{\varepsilon})^2 \sum_{\substack{i=1 \\ i > j}}^{N} \sum_{j=1}^{N} \left(\frac{\partial^2 J}{\partial c_i \partial c_j} \bigg|_{\infty} \right)^2. \tag{6.11}$$

Recall the application of the central limit theorem in section 6.1. We can make the same assumption for our higher-order statistical wordlength derivation. For the usual digital filtering estimate, the coefficient quantization could either decrease or increase the (zero-mean) error in the transfer function magnitude at any specific frequency. In the control case however, the value of J can only *increase* under coefficient quantization. Thus we only need a specification on the maximum allowed value of J including the degradation due to coefficient quantization: $J_\infty + E_0$. Following the general approach of section 6.1, this value can be related to the two-sigma point in the distribution for dJ (see figure 6.1):

$$J_\infty + E_0 = J_\infty + \overline{dJ} + 2\sigma_{dJ}. \tag{6.12}$$

This choice of σ_{dJ} gives a 97.5% confidence level in terms of remaining below the allowed deviation E_0. Combining (6.11), (6.12), and the values of $\overline{\varepsilon}$ and σ_ε, we can derive an expression for Δ^2:

$$\frac{1}{\Delta^2} = \frac{1}{24E_0} \sum_{i=1}^{N} \left(\frac{\partial^2 J}{\partial c_i^2}\bigg|_x \right)$$

$$+ \frac{1}{6E_0} \left[\sum_{\substack{i=1 \\ i>j}}^{N} \sum_{j=1}^{N} \left(\frac{\partial^2 J}{\partial c_i \, \partial c_j}\bigg|_x \right)^2 + \frac{1}{5} \sum_{i=1}^{N} \left(\frac{\partial^2 J}{\partial c_i^2}\bigg|_x \right)^2 \right]^{1/2}. \tag{6.13}$$

Using (6.6), the *SWL* can be written

$$SWL = l + \frac{1}{2} \log_2 \left(\frac{1}{\Delta^2} \right). \tag{6.14}$$

There are several important distinctions between the statistical word-length method as described in section 6.1 and expression (6.14). First, the requirement of second-derivative terms has led to a fairly complex expression for the *SWL*; thus an efficient computational procedure will be critical. Second, since the performance index J is not frequency dependent, neither is the statistical wordlength estimate. Only one evaluation will be needed, rather than one per frequency as with the transfer function-based filter wordlength estimate. Another distinction involves the

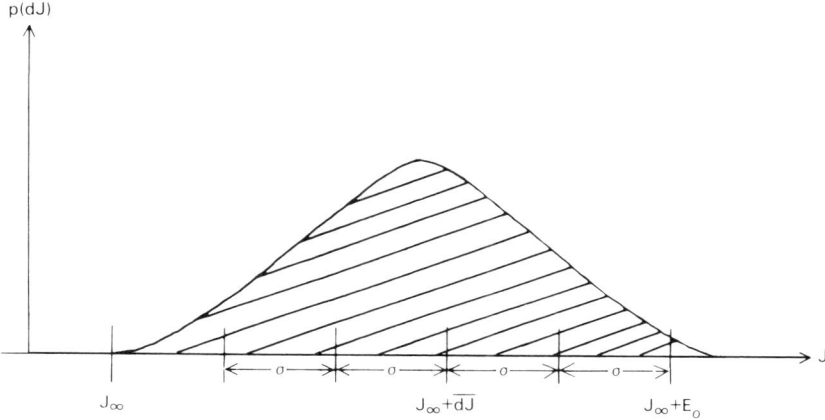

Figure **6.1** Probability density of *dJ*.

Gaussian distribution assumption. The analysis in (6.12)–(6.14) applied the central limit theorem to justify this distribution. Yet we know that the distribution of dJ must be somewhat skewed, since dJ is always positive. Thus the 97.5% probability figure may not be as accurate as the digital filter probability of 95%, given the same number of summed uncorrelated terms N. However, the exact probability level is not important, as long as it is relatively high.

The final distinction between the usual filtering estimate described in section 6.1 and the LQG controller estimate is the nonzero mean degradation $\overline{dJ}$. For the filtering case, the mean degradation in the transfer function is zero. For the LQG development, it is possible to form an estimate without taking into account the standard deviation of the error dJ. If we set the mean degradation value to equal the allowed degradation E_0, then, using (6.9),

$$\overline{dJ} = E_0 = \frac{\Delta^2}{24} \sum_{i=1}^{N} \left(\frac{\partial^2 J}{\partial c_i^2} \bigg|_\infty \right). \tag{6.15}$$

From (6.15), we can write an expression for Δ^2:

$$\Delta^2 = \frac{24E_0}{\sum_{i=1}^{N} \left(\frac{\partial^2 J}{\partial c_i^2} \bigg|_\infty \right)}. \tag{6.16}$$

A *mean* statistical wordlength ($MSWL$) can now be defined, using (6.14) and (6.16):

$$MSWL = l + \frac{1}{2}\log_2 \left[\frac{\sum_{i=1}^{N} \left(\frac{\partial^2 J}{\partial c_i^2} \bigg|_\infty \right)}{24E_0} \right]. \tag{6.17}$$

The interpretation of this estimate is as follows: If we use the $MSWL$ estimate as the actual wordlength, half the structures in the ensemble will have more degradation than E_0 and half will have less. The usefulness of the $MSWL$ estimate depends on how much we can infer about a reliable estimate (one that is correct in most cases) of the required wordlength. Consequently, it will be a function of the width of the dJ distribution in figure 6.1 and of the change in this distribution from structure to structure. In other words, it will depend on the tightness in the relation between the SWL and the $MSWL$ estimates. As a test, we shall compute both estimates for a selection of structures. If we *can* use the $MSWL$, its advantage is clear—reduced complexity and, it is hoped, significantly less computation time.

At this point it is convenient to consider the analysis of suboptimal compensators. If a suboptimal compensator results from a parameter optimization problem [75, 76, for example], then the first-order sensitivities will all be zero and the statistical wordlength approach developed in this chapter can be used. If the suboptimal design is only an approximation of an optimal design, then we can still apply this method; the only difference would be the inclusion of first-derivative $(\partial J/\partial c_i)|_{\infty}$ terms (first-order sensitivities). These terms would be nonzero since the compensator is not optimal, or even locally optimal as in the parameter optimization designs.

6.3 Computing the Statistical Wordlength

As in chapter 5, the trace form of J will be convenient for computing statistical wordlength. Recall the following two equations from chapter 5:

$$J = \text{trace}(\tilde{\Upsilon}\tilde{Z}), \tag{6.18}$$

$$\tilde{Z} = \tilde{A}\tilde{Z}\tilde{A}' + \tilde{C}, \tag{6.19}$$

where $\tilde{\Upsilon}$ contains the weighting matrices Q, M, and R as in (5.35), and $\tilde{A}$ and $\tilde{C}$ are defined by

$$\tilde{A} = \left[\begin{array}{c|c|c} \Phi & 0_n & \Gamma\rho \\ \hline \Psi_{12}L & & \Psi_{11} \end{array}\right]$$

and

$$\tilde{C} = \begin{bmatrix} \Theta_1 & 0 \\ 0 & \{\Psi_{12}\Theta_2\Psi'_{12}\} \end{bmatrix}.$$

For the remainder of this chapter, assume that we are always dealing with scaled compensators; thus the tilde superscript will be dropped.

The matrices Ψ_{11} and Ψ_{12} are derived from the infinite-precision *scaled* compensator parameters. If (6.18) is evaluated with these infinite-precision matrices, the resulting value of the performance index J_{∞} will be independent of the structure chosen. However, the partial derivatives of J_{∞} with respect to the coefficients of the structure, evaluated at the ideal coefficient values, *will* be structure dependent since each coefficient c_i resides in a particular Ψ matrix. The second partial derivative of (6.18) can be written

$$\frac{\partial^2 J}{\partial c_i \partial c_j} = \text{trace} \left[\Upsilon \frac{\partial^2 Z}{\partial c_i \partial c_j} \right]. \tag{6.20}$$

Thus the partial derivatives of Z (each a matrix) must be computed efficiently. Taking the first derivative with respect to c_i of (6.19) produces

$$\frac{\partial Z}{\partial c_i} = A \frac{\partial Z}{\partial c_i} A' + Q_i + Q_i' \tag{6.21}$$

where

$$Q_i = \frac{\partial A}{\partial c_i} ZA' + \begin{bmatrix} 0 & 0 \\ 0 & \left\{ \dfrac{\partial \Psi_{12}}{\partial c_i} \Theta_2 \Psi_{12}' \right\} \end{bmatrix}.$$

The second partial derivatives of Z can now be written

$$\frac{\partial^2 Z}{\partial c_i \partial c_j} = A \frac{\partial^2 Z}{\partial c_i \partial c_j} A' + X_{ij} + X_{ij}', \tag{6.22}$$

where

$$X_{ij} = \frac{\partial A}{\partial c_j} \frac{\partial Z}{\partial c_i} A' + \frac{\partial A}{\partial c_i} \frac{\partial Z}{\partial c_j} A' + \frac{\partial A}{\partial c_i} Z \frac{\partial A'}{\partial c_j} + \frac{\partial^2 A}{\partial c_i \partial c_j} ZA'$$

$$+ \begin{bmatrix} 0 & 0 \\ 0 & \left\{ \dfrac{\partial \Psi_{12}}{\partial c_i} \Theta_2 \dfrac{\partial \Psi_{12}'}{\partial c_j} + \dfrac{\partial^2 \Psi_{12}}{\partial c_i \partial c_j} \Theta_2 \Psi_{12}' \right\} \end{bmatrix}.$$

Rather than solving (6.22) N^2 times, which is extremely time consuming, we can apply the adjoint method used in chapter 5. Equations (6.20) and (6.22) can be replaced by the following two equations:

$$\frac{\partial^2 J}{\partial c_i \partial c_j} = \text{trace } \hat{U}(X_{ij} + X_{ij}') = 2 \, \text{trace}(\hat{U} X_{ij}), \tag{6.23}$$

$$\hat{U} = A' \hat{U} A + \Upsilon, \tag{6.24}$$

where $\hat{U}$, A, and Υ are all $(2n + 1) \times (2n + 1)$ matrices and A and Υ can be found in (6.19) and (5.35).

Further simplification is possible when evaluating (6.23) once $\hat{U}$ is computed. The matrices A and $\Psi_{12}\Theta_2\Psi'_{12}$ can be expressed in terms of Ψ_∞:

$$
A = \begin{bmatrix} I_n & 0 \\ 0 & \Psi_\infty \end{bmatrix}
\begin{bmatrix} \Phi & 0_n & \Gamma_\rho \\ \hline 0 & I_{n+1} & \\ \hline L & 0 & \end{bmatrix},
\tag{6.25}
$$

$$
\Psi_{12}\Theta_2\Psi'_{12} = \begin{bmatrix} I_n & 0 \\ 0 & \Psi_\infty \end{bmatrix}
\begin{bmatrix} 0_n & 0 & 0 \\ 0 & 0_{n+1} & 0 \\ 0 & 0 & \Theta_2 \end{bmatrix}
\begin{bmatrix} I_n & 0 \\ 0 & \Psi'_\infty \end{bmatrix}.
\tag{6.26}
$$

Thus the expression for X_{ij} can be grouped into four terms based on derivative quantities: one involving $\partial\Psi_\infty/\partial c_j$ and $\partial Z/\partial c_i$, one involving $\partial\Psi_\infty/\partial c_i$ and $\partial Z/\partial c_j$, a third involving $(\partial^2\Psi_\infty)/(\partial c_i\partial c_j)$, and the last involving $\partial\Psi_\infty/\partial c_i$ and $\partial\Psi_\infty/\partial c_j$. The first partial derivatives of Z are known from solving (6.21) N times. Now let us examine the derivatives of Ψ_∞. Since a coefficient can occur once in only one Ψ_m matrix, the first partial of Ψ_∞ with respect to c_i [assume c_i is located at index (k,l) in Ψ_m] will be

$$
\frac{\partial\Psi_\infty}{\partial c_i} = \Psi_q\Psi_{q-1}\cdots\Psi_{m+1}E_{kl}\Psi_{m-1}\cdots\Psi_1,
\tag{6.27}
$$

where E_{kl} is defined to be a *unit element matrix* of the same dimensions as Ψ_m. This matrix is all zero except for a single unity entry at index (k,l). Similarly, if c_j is located in Ψ_t at index (r,s), we can write

$$
\frac{\partial^2\Psi_\infty}{\partial c_i\partial c_j} = \Psi_q\cdots\Psi_{m+1}E_{kl}\Psi_{m-1}\cdots\Psi_{t+1}E_{rs}\Psi_{t-1}\cdots\Psi_1.
\tag{6.28}
$$

We can infer from (6.28) that if c_i and c_j are in the same matrix Ψ_m (thus $m = t$), then $(\partial^2\Psi_\infty)/(\partial c_i^2)$ must be zero. This fact simplifies the calculation of X_{ij} to some extent (more significant for the $MSWL$ estimate, which only requires X_{ij} for $i = j$). Appendix C presents further details regarding the evaluation of (6.23).

Unfortunately, the evaluation of (6.23) still requires the computation of equation (6.21) for all N coefficients. However, these computations can also be simplified. The Lyapunov solution method used in this analysis is that of Barraud [77] and has several distinct parts. Given an equation like (6.19), this method will

1. compute an orthogonal transformation matrix P that converts A to *upper Schur form* (upper triangular except for the first subdiagonal row): $A_s = P'AP$;

2. use P to transform C to C_s: $C_s = P'AP$;

3. solve the transformed equation $Z_s = A_s Z_s A_s' + C_s$ by a back substitution technique;

4. transform the result Z_s to Z via $Z = PZ_s P'$.

The number of operations involved in each step is proportional to $(2n + 1)^3$ if Z, A, and C are $(2n + 1) \times (2n + 1)$. However, by far the majority of the computations are involved in step 1, which performs an eigenvalue-eigenvector analysis of A. Step 3 requires (approximately) 5 to 10% of the total time, depending on the particular A matrix. The important point to realize is that step 1 need only be performed once for all N equations (6.21). In fact, steps 2 and 4 can also be simplified by including the P and P' multiplications in the precomputed matrices $M1$ and $M2$ described in appendix C. Using this method, there will still be a proportionality to $N(2n + 1)^3$ in computing (6.21), but it will be many times smaller than for the full four-step procedure.

In summary, the computational procedure for statistical wordlength primarily involves the second derivatives of J required for (6.10). Assuming that computation time is dominated by the number of multiplies, the following approximate dependence of the computation time on the number of coefficients N and the (augmented) system order $2n + 1$ exists:

$$t_{SWL} \propto N^2(2n + 1)^2 + N(2n + 1)^3 + (2n + 1)^3. \qquad (6.29)$$

For the $MSWL$ estimate, this proportionality will be reduced:

$$t_{MSWL} \propto N(2n + 1)^3 + (2n + 1)^3. \qquad (6.30)$$

Thus as N increases, the $MSWL$ estimate becomes computationally more and more efficient as compared with the SWL estimate.

6.4 Direct Wordlength Computation

For comparison, it is important to include a direct method for determining the coefficient wordlength required to meet or exceed the degradation level E_0. Basically, such a procedure will involve selecting a test wordlength, rounding the coefficients to that wordlength, and then forming the (finite-precision) matrices Ψ_i for $i = 1, \ldots, q$, A, and C. Using these finite-precision parameters, the Lyapunov equation (6.19) must be solved and the trace (6.18) evaluated. The resulting value of J can be compared to $J_\infty + E_0$, and then a decision made whether to alter the test wordlength up or down.

If the performance index were strictly monotonic in the coefficient wordlength, then a binary search algorithm could be designed that would always succeed in finding the required wordlength. For example, starting at some large initial test wordlength, one could decrement the test word-length 10 bits at a time until the performance index exceeded the value $J_\infty + E_0$, then increment the test wordlength in smaller steps until the performance index was below $J_\infty + E_0$, and so forth. However, J need not be strictly monotonic in wordlength; the coefficient rounding operation is quite nonlinear. However, J is *roughly* monotonic. Thus the search procedure must try to account for possible anomalies in the behavior of J. One other pitfall must be avoided: If the test wordlength is so small that the resulting feedback system is *unstable*, then the computed J value will be meaningless. One simple way to test for this possibility would be to examine the resulting eigenvalues, which are a by-product of the Lyapunov solution method of Barraud.

The method we have used is based on the above discussion. After loosely bracketing the allowed degradation E_0 with two test wordlengths, the lower of which is tested to guarantee stability, fit an exponential curve to these two points. Using E_0 and this curve, a reasonable choice of a new test wordlength can be made. From this point, the test wordlength is stepped a bit at a time until the required wordlength is established. The details of the algorithm are as follows:

1. Bracket the wordlength w with the initial values $w_{max} = 48$ and $w_{min} = 0$. Initialize the increment i to 10, and set the initial value of w near the value w_{max}. Compute the ideal J_∞ using the double-precision coefficient values, and add an allowed level of degradation to produce the desired performance J_0.

2. Decrement the wordlength w by i.

3. Test for a negative wordlength w. If found, set w to 1.

4. Round the ideal coefficient values to the wordlength w, and compute the resulting test value J_t of the performance index.

5. Test for instability by comparing J_t to J_∞. If J_t is smaller, the system

with coefficients of wordlength w is unstable. Then increment w by i, halve the increment size, and return to step (2). Otherwise, if J_t is larger than J_∞, proceed to the next step.

6. Test to see whether J_t is between J_∞ and J_0. If so, set w_{max} to the current value of w and return to step (2). Otherwise, set w_{min} to the current value of w and proceed to the next step. Thus we have bracketed the required wordlength with w_{max} and w_{min} and know the performance levels for each of these wordlengths.

7. Using the two wordlength-performance points found in step (6), and also the ideal performance value J_∞ (associated with some very large wordlength, say 100), fit an exponential curve to describe the performance index as a function of the wordlength. Interpolate to find a next guess at the required wordlength. Round the coefficients to this wordlength and compute the resulting performance index.

8. If this value is greater than J_0, increase w a bit at a time until the resulting performance level is below J_0. The corresponding w will be the required wordlength. If, however, the performance level from step (7) is below J_0, decrease w a bit at a time until the resulting performance level is above J_0. The corresponding wordlength w, *plus* one bit, will be the required wordlength.

The direct algorithm may be time consuming as compared to the statistical method because it requires repeated solutions of the Lyapunov equation (6.19) until we determine the required wordlength and, further, no simplifications are possible from one solution to the next since each finite-precision A matrix is different. If an average of n_i iterations are required to establish the required wordlength, then the dominant number of multiply operations required to compute this *true wordlength* (*TWL*) has the following proportionality:

$$t_{TWL} \propto n_i (2n + 1)^3. \tag{6.31}$$

Thus a comparison between the statistical estimates *SWL* and *MSWL* and the *TWL* just described will depend upon n, n_i, N, and the constants of proportionality. However, as the number of coefficients increases, the statistical estimates will become less and less efficient, while the true wordlength computation time will remain essentially constant. Recall, however, that the statistical estimate can be used as the basis for a wordlength optimization procedure (see chapter 8). The true wordlength method could not be used for such a procedure since it is not continuous and thus not differentiable.

6.5 The F8 System and the Coefficient Wordlength Issue

The effects of finite-coefficient wordlength were evaluated for the F8 system example and the 10 structures described in chapter 5. The results are presented in table 6.1 using the following format. Column 1 lists the number of integer bits required for the coefficient word; this value is obtained from the largest scaled coefficient value (see table 5.2). The next three columns list the statistical estimates SWL and $MSWL$, and finally the true wordlength TWL as evaluated in section 6.4. The execution time in seconds for each wordlength determination method is listed in parentheses following each entry. These times are subject to some small amount of uncertainty depending on specific run-time conditions, so they must be regarded as approximate. As in chapter 5, the wordlengths listed represent the number of coefficient bits (not including the sign) required to achieve at most a 5% increase in the performance index J. Finally, the last column of table 6.1 lists the number of bits by which the SWL estimate exceeds the actual required wordlength.

A great deal of information may be drawn from table 6.1. First, we can discuss the performance of the 10 structures with regard to coefficient wordlength. Referring to the TWL values, we can see that the parallel structure (b) using first-order and second-order direct form II sections and the block optimal parallel structure (f) performed the best, needing

Table **6.1** F8 coefficient wordlength results

Structure	1	SWL	$MSWL$	TWL	SWL-TWL
(a) Direct form II	16	35.99 (0.81)	35.05 (0.70)	32 (1.2)	3.99
(b) Parallel d.f. II	1	6.84 (0.93)	6.16 (0.78)	6 (1.08)	0.84
(c) Parallel d.f. II	4	12.38 (0.87)	11.52 (0.78)	11 (1.26)	1.38
(d) Parallel d.f. II	4	19.02 (0.85)	18.14 (0.77)	13 (1.08)	6.02
(e) 1-Level from (c)	3	11.08 (0.90)	10.22 (0.78)	10 (1.19)	1.08
(f) Block optimal	1	7.02 (1.26)	6.2 (0.91)	7 (1.11)	0.02
(g) Cascade, d.f. II	11	26.25 (0.83)	25.38 (0.72)	21 (1.21)	5.25
(h) Cascade, d.f. II	6	14.61 (0.86)	13.81 (0.72)	14 (1.36)	0.61
(i) Cascade, d.f. I	9	24.25 (0.84)	23.38 (0.71)	20 (1.1)	4.25
(j) Simple	1	9.05 (2.44)	8.25 (1.29)	9 (1.71)	0.05

only 6 and 7 bits, respectively. Quite acceptable performance was also achieved with the simple structure (j) (9 bits), the one-level parallel structure (e) (10 bits), and the parallel structure (c) (11 bits). As with the roundoff noise results of chapter 5, the direct form II structure (a) performed the worst. For the two parallel and two cascade structures using all second-order sections but with two different pole pairings, the pairing that was better for roundoff noise [structures (c) and (h)] was also superior for coefficient sensitivity—2 bits better for the parallel case, and 7 bits better for the cascade.

In fact, if we rank the structures on the basis of their required coefficient wordlengths (b, f, j, e, c, d, h, i, g, a) and then also on the basis of their signal variable wordlengths for fixed roundoff noise performance (f, b, j, e, c, h, d, i, g, a), we can see a very strong correlation. The orderings are nearly identical—only the adjacent structures (b) and (f) are interchanged, as are (h) and (d). The correlation between good roundoff noise performance and low coefficient sensitivity has been well publicized for digital filter structures [17, 42, 50, 78, 79]. Of course, these results pertain to the sensitivity of the transfer function *magnitude* to its coefficients. From our results, this correlation seems to carry directly over to the LQG control compensator.

One point to be cognizant of is that certain coefficients in a structure, when rounded to the *TWL* wordlength, may in fact become zero or unity, thus eliminating them as multipliers. This situation occurs in the simple structure (j), reducing the number of coefficients from 50 to 40; in the block optimal structure (f), reducing the number of coefficients from 25 to 24; and in the parallel structure (b), reducing the number of coefficients from 17 to 16. Such reductions should factor into the structure selection procedure.

Taking the number of multiplies, number of precedence levels, roundoff noise performance, and required coefficient wordlength all into account, the parallel structure (b), which uses first-order sections for real poles and second-order sections for complex poles, is probably the best choice. To achieve an *overall* 3% maximum increase in J (including roundoff noise and coefficient rounding effects) with this structure, we could use an 8-bit A/D converter, 8-bit coefficients, and 10-bit signal variables. (Due to the quadratic nature of J, each extra bit reduces the increase in J by approximately a factor of four.) Each of these wordlengths includes the sign bit. If circumstances *required* a one-level structure for a short sampling period T, then we would probably use the block optimal structure and 24 hardware multipliers. Any final decision as to structure selection is of course application dependent.

This discussion applies to the actual wordlengths found by the direct method. Now let us examine how useful it would be to make the comparison of structures using the SWL statistical estimate. For the 10 structures shown, the SWL estimate was conservative by up to 6 bits, which is quite a wide range. However, this situation is easily explained. Structures (d), (g), and (i) had the poorest estimates. Not coincidentally, all three of these structures have two particular coefficients in common, -0.9938344 and 1.9938281 (see appendix A), and these two coefficients dominate in the expression for statistical coefficient wordlength for these examples. Removal of these two coefficients from the statistical word-length analysis produces new estimates within *one* bit of the true word-length. Thus this problem probably represents a case in which we have violated the assumption of independent coefficient errors; due to the very high degree of similarity in the fractional portions of the two coefficients, their errors must be correlated in such a way that they cancel. It is also possible that this situation represents a low probability event (in the ensemble sense); in other words, this case lies in the left-hand tail of the distribution of figure 6.1. In any case, these particular two coefficients resulted from pairing the two real near-unit-magnitude poles, which has already been shown to be a poor choice with respect to finite wordlength performance (see chapter 5). Perhaps of more importance, over the 10 structures, the SWL estimate is excellent (conservative by at most 1.4 bits) for the five lowest coefficient wordlength structures and the cascade (h).

As for a comparison between the SWL and $MSWL$ estimates, the $MSWL$ value was consistently 0.68–0.94 bits below the SWL value. This tight range of values suggests that the distribution of dJ shown in figure 6.1 is quite narrow. Thus the $MSWL$, which is simpler to compute, may well be preferable to the SWL. One could compute the $MSWL$ and then add some fixed number, say one bit, for an estimate. The primary advantage to using the $MSWL$ estimate over the SWL, given their apparent tight correlation, would be in the constrained optimization procedure of chapter 8. In principle, the optimization procedure could use either statistical estimate as its objective function. However, the simpler $MSWL$ estimate leads to a less complex optimization procedure. In chapter 8 this estimate will be used as the basis for finding a minimum coefficient wordlength structure.

Figure 6.2 plots approximate execution times of the *TWL*, *SWL*, and *MSWL* routines, as run on an Amdahl 470 at the Charles Stark Draper Laboratory, versus the number of coefficients in the structure. From this figure we can see that the *TWL* computation takes between 1.1 and 1.35 seconds [since the routine could be written more efficiently to reduce the execution time for structure (j)]. For approximately 20 coefficients or less, the *SWL* estimate is somewhat faster to compute than the *TWL* value, about 15 to 30%, and the *MSWL* is at least another tenth of a second faster than this. However, keep in mind that the primary advantage to a statistical wordlength estimate is the structural optimization procedure that can be derived from it, and not its computational speed.

Another important advantage to either the *SWL* or *MSWL* estimates

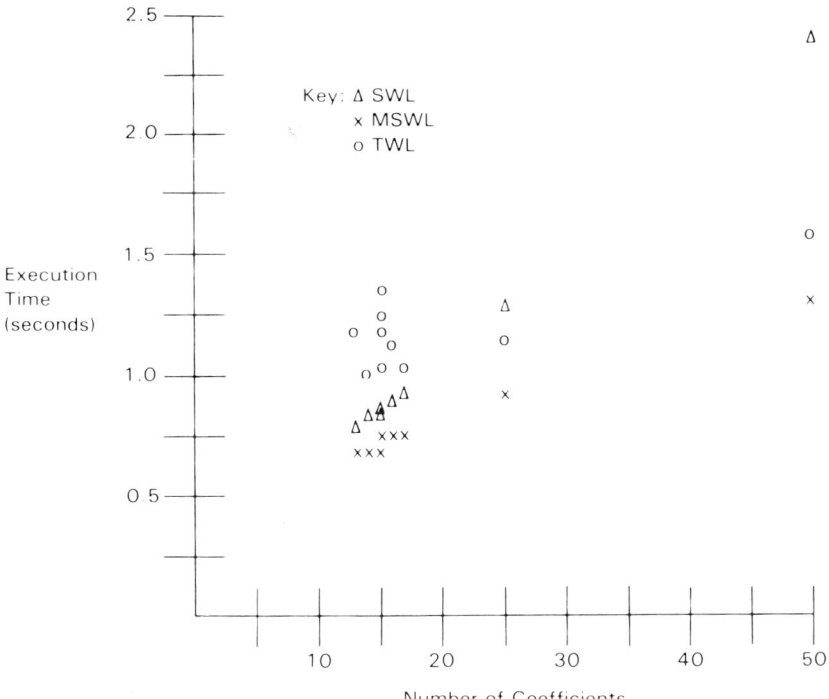

Figure **6.2** Execution times versus number of coefficients.

is the second-order sensitivities they produce. Once these values are computed, it is easy to see which coefficients dominate, as far as the required coefficient wordlength is concerned. The portion of the structure in which these coefficients occur is then a likely candidate for optimization as described in chapter 8. For example, if we consider a cascade structure composed of second-order sections, then the section which has the most sensitive coefficients should be the section that is optimized. In other words, we should allow the optimization procedure to exploit any possible degrees of freedom in that section. This might result in a few more coefficient branches, but it will probably lead to an overall structure with substantially fewer required coefficient bits.

In addition, there is a further advantage to knowing the individual sensitivities. In table 6.1 we can see that structures (a), (c), (d), (e), (g), (h), and (i) have at least one large coefficient that requires the large number of integer bits (more than 1) in the fixed-point coefficient word. By replacing each of these coefficients by a smaller-magnitude coefficient followed by a shift, we can reduce the number of integer bits that are required. The amount of the shift (the number of bits) will be limited by the coefficient sensitivities. For example, structure (d) has only 2 coefficients larger than 2 (see appendix A). Their ideal values are approximately 15.7 and -15.7. From the SWL analysis, their sensitivities $(\partial^2 J)/(\partial c_i^2)$ are approximately 0.043. The dominant sensitivities with respect to determining the actual coefficient wordlength (and for the sake of this discussion we will leave out the coefficients 1.99383281 and -0.9938344 previously mentioned) are on the order of 150. Since each decrease by a factor of 2 in a coefficient value results in an increase by a factor of 4 in its sensitivity (because we are taking second-order sensitivities), we can decrease these two large coefficients by a factor of 8 (3 bits), while only increasing their sensitivities to about 2.8. Since this is still insignificant with respect to 150, the statistical (and true) fractional wordlengths will not increase appreciably. The net result is a savings in total wordlength of 3 bits (from 13 to 10 total bits), while adding only two simple 3-bit shifts to the hardware. Note that such a shift operation does not involve any additional hardware, but just a rewiring of the respective multiplier output and the following quantizer or adder input. In structure (d), all we are doing is replacing a multiplication by 15.7173777648272 (before rounding) with a multiplication by 1.9646722206034 and a 3-bit shift (a multiplication by 8) and similarly for the other large coefficient. Table 6.2 shows the

Table **6.2** Shifting to reduce coefficient wordlength

Structure	l	TWL (no shifts)	Possible shift (bits)	Expected TWL (with shifts)
(a) Direct form II	16	32	11	21
(b) Parallel direct form II	1	6	Unnecessary	6
(c) Parallel direct form II	4	11	3	8
(d) Parallel direct form II	4	13	3	10
(e) Parallel 1-level from (c)	3	10	2	8
(f) Block optimal parallel	1	7	Unnecessary	7
(g) Cascade, direct form II	11	21	6	15
(h) Cascade, direct form II	6	14	3	11
(i) Cascade, direct form I	9	20	4	16
(j) Simple	1	9	Unnecessary	9

reduction possible for all 10 structures (where this method applies). Structures (c) and (e) now are so much better in terms of required wordlength that they are nearly as good as the best choices (b) or (f).

6.6 Joint Analysis of Roundoff Noise and Coefficient Rounding Effects

Chapters 5 and 6 have presented analyses of roundoff noise effects and finite coefficient wordlength effects as if the two were completely independent. Ideally though, one would want to analyze the roundoff effects on a structure using its actual finite-wordlength coefficients. However, the structure must of course be scaled *before* the coefficient wordlength analysis can be carried out. Thus, a complete structural evaluation procedure would (1) scale the structure, (2) compute its required coefficient wordlength, (3) round the infinite-precision coefficient values to that wordlength, and finally (4) compute the necessary signal variable wordlength via a roundoff noise analysis using the rounded coefficient values. The procedure we have followed differs in that the roundoff analysis is performed using the infinite-precision coefficient values, rather than the rounded values. This simplification was made for two reasons. First, the effect of using infinite-precision coefficients in the roundoff analysis causes only very minor changes as compared to using the finite wordlength coefficients. (It is essentially a second-order effect.) Second, the nature of the roundoff analysis procedure is *approximate* to start with. We are adding just one more small approximation. Once the roundoff analysis procedure of chapter 5 and the statistical coefficient wordlength determination methods of chapter 6 are used to select one from a group of candidate structures, then it would be advisable to go back and do a more careful analysis of the finite-wordlength effects and required wordlengths for this structure.

A more important observation is the following: It has been assumed that the increase in J due to roundoff noise (including the A/D contribution) must be limited to some level, for example, 5% of the ideal J, and that the increment due to finite wordlength coefficients must also be limited to some level E_0, for example, also 5%. Thus the total degradation will be approximately the sum of these values, or 10%. However, there is no implicit reason why the overall error budget must be split evenly between these two effects. In fact, once a structure is selected using the techniques described in chapters 5 and 6, the respective required wordlengths can be modified, perhaps to *convenient* or *more nearly equal* values, by apportioning the two error limitations differently. Such a degree of freedom can be exploited to help simplify the hardware by conforming to more standard wordlengths and thus less expensive and more available hardware components.

6.7 Summary

This chapter has examined the coefficient wordlength issue for digital feedback compensators, assuming fixed-point arithmetic. We have described several methods for determining the required coefficient wordlength of a structure and for selecting structures with low required wordlengths and have stressed the statistical approach first used for digital filters. This statistical approach results in a wordlength estimate that can be shown to be inadequate for LQG compensators due to the optimal nature of an LQG design. Through the inclusion of second-order sensitivities in the statistical formulation, we have derived a statistical estimate that is appropriate to the LQG problem, and in fact to any design problem involving the optimization of a scalar performance criterion. We have also described an efficient computational technique for evaluating this estimate. As a comparison, we have presented a direct method for determining the required coefficient wordlength and evaluated both techniques for 10 example structures.

Based on the results presented in section 6.5, it may be concluded that the statistical estimates SWL and $MSWL$ are not simple enough to justify their use (instead of the direct TWL calculation) on a calculation-time basis *alone*. However, they have two excellent advantages that overwhelmingly recommend their use. First, the resulting second-order sensitivities are an excellent guideline for (1) reducing the required word-length of certain structures with large coefficients (those greater than two), and (2) discovering which sections of a structure dominate in determining the required wordlength (this information could be used to select which portion of a structure to optimize, as discussed in chapter 8). Second, through the use of the statistical wordlength measure as an objective function, we can effectively synthesize constrained minimum coefficient wordlength structures by applying transformations as described in chapter 8. Once a set of candidate structures has been compared with regard to their roundoff noise, coefficient wordlength effects (using the statistical estimates), precedence levels, and so forth, and a structure selected, it should be analyzed in more detail. Specifically, it would definitely be worthwhile to evaluate the *TWL* as a final step in determining the required coefficient wordlength.

7

Finite-Wordlength Effects: Limit Cycles

The roundoff noise analysis presented in chapter 5 relies on the validity of the additive white noise model for roundoff quantization. In this chapter we shall address the cases for which this model is not valid. In particular, a digital structure can exhibit oscillations known as *limit cycles*. Essentially, any linear system including one or more nonlinearities in a feedback loop can exhibit autonomous oscillations due to those nonlinearities. For digital filters or compensators, quantization nonlinearities exist after each multiplication product or sum of products, and overflow nonlinearities exist after each adder. In addition, both nonlinearities exist at the A/D converter. We can classify the resulting oscillations as quantizer limit cycles or overflow limit cycles, depending on the type of nonlinearity that causes them. Of these two types, the overflow limit cycle tends to be more serious in its effect on performance: when it occurs, it has an amplitude equal to the maximum representable digital signal.

In the digital signal processing literature, there is a great number of results concerning limit cycles. An excellent review of this literature on limit cycles can be found in Kaiser [22], or in the finite wordlength survey articles by Classen, Mecklenbräuker, and Peek [62] and Oppenheim and Weinstein [59]. Willsky [16] compares these results to the nonlinear system stability results known to the control and estimation field. Rather than cataloging all the different results and techniques used for dealing with limit cycles in digital filters, this effort will be confined to the more general approaches, since they are more likely to extend to the control environment.

Several points concerning the digital signal processing limit cycle results should be mentioned. First, most of these results concern zero-input limit cycles, oscillations that occur when there is no input driving the filter. When a nonzero input *is* present, it is unclear just what limit cycle behavior means, since the response of the filter to the input can be superimposed on an oscillation or can actually *eliminate* the oscillation [80]. Second, most of the digital filtering limit cycle results are specific to a single structure, usually the second-order direct form II structure. Since limit cycles can only be caused by nonlinearities in the recursive part of a filter, these results are further specific to the pole section of the direct form II structure. Two general conclusions follow from the digital filtering results. First, for avoiding quantizer limit cycles, sign-magnitude truncation is to be preferred over roundoff. Recall that the reverse is true when quantization noise minimization is considered. Second, for avoiding overflow limit cycles, the saturation characteristic is to be preferred over the two's complement overflow characteristic. For overflow, it is important to keep in mind that the two's complement characteristic requires no additional hardware—it is implicit in any addition using two's complement arithmetic. Additional hardware is required to implement the saturation characteristic.

As a whole, our results concerning limit cycles in digital feedback compensators are limited. However, in this chapter we shall make four observations. First, we shall point out that zero-input limit cycles *always* occur for control systems with open-loop unstable plants. Second, we shall stress just how the feedback loop of a control system can alter the limit cycle performance of a digital compensator. In fact, even if the compensator *alone* has no limit cycles, the feedback system of plant and compensator together *can* exhibit limit cycles. Third, for a variety of reasons, we shall argue that the limit cycle results in digital signal processing do not generally apply to the control setting. Finally, we shall discuss the significant question of whether limit cycles themselves are an issue at all for LQG systems. At even the simplest level, no LQG system could be thought of as zero-input, given the system driving and measurement noises.

The remainder of this chapter is organized as follows. In sections 7.1 and 7.2 we shall present the more general approaches used in digital signal processing for dealing with quantizer limit cycles and overflow limit cycles, respectively. Then, in section 7.3 we shall consider the various aspects of the limit cycle problem as it concerns digital feedback compensators. Specifically, the approaches described in 7.1 and 7.2 will be dealt with in greater depth.

7.1 Quantizer Limit Cycles

There are three basic methods for dealing with the limit cycles caused by the quantization nonlinearities in a digital structure. The first of these is simply to apply general *nonexistence* results, which guarantee that limit cycles do not occur. Many of these are so general as to apply to the overflow case as well. This procedure can be quite restrictive as to the types of structures and quantizers (roundoff or sign-magnitude truncation) that apply. The second approach is quite different; if we can bound the magnitude of the quantization effects (this bound would include limit cycle *and* noise effects) to some level dependent on the signal variable wordlength, then we need only use wordlengths long enough to make these effects negligible. Such analysis techniques are frequently based on Lyapunov theory [16]. Finally, the last procedure involves *random rounding*; basically this refers to the technique of adding randomness at selected points in a structure to break up potential limit cycles. Of course this technique tends to add noise to the system, which requires longer wordlengths to restore performance to desired levels.

7.1.1 General Nonexistence Results

Two general nonexistence results described in the digital signal processing literature are of interest. The first of these is a frequency-domain criterion introduced by Claasen, Mecklenbräuker, and Peek [81] and based on the *sector* nature of the quantizer and/or overflow nonlinearities. Let us divide the digital filter under consideration, which generally will have *multiple* nonlinearities, into its linear and nonlinear portions, as in figure 7.1. The signals $\varepsilon_i(k)$ and $g_i(k)$ will represent the input and output of the ith nonlinearity. The linear portion of the system in figure 7.1 will be described by the transfer response matrix $W(z)$, where

$$\zeta(z) = W(z)G(z) \tag{7.1}$$

and $\zeta(z)$ and $G(z)$ are the z-transforms of $\varepsilon(k)$ and $g(k)$, respectively.

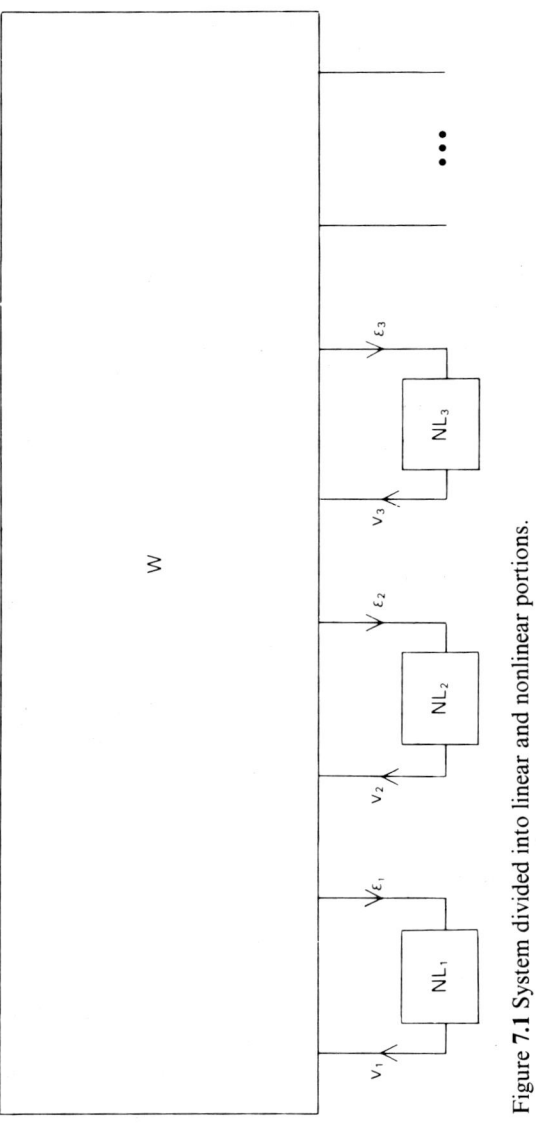

Figure **7.1** System divided into linear and nonlinear portions.

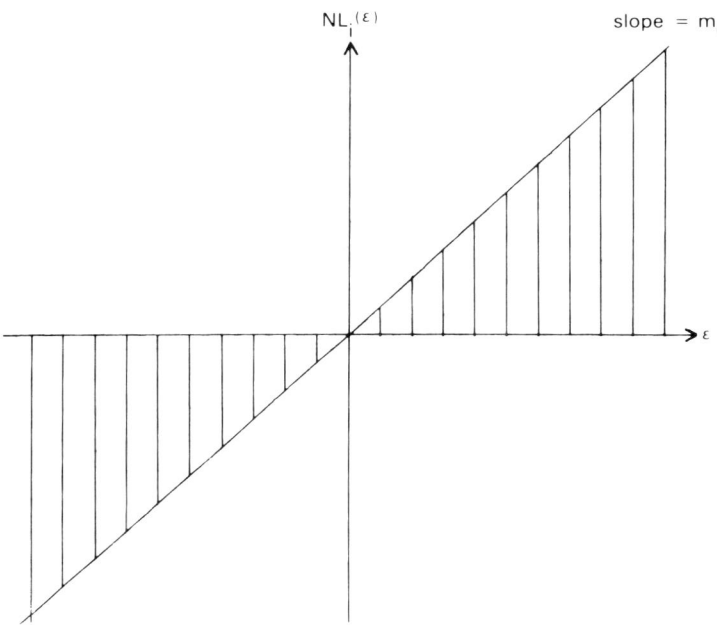

Figure **7.2** Sector nonlinearity.

Now let us assume that the ith nonlinearity is a *sector* nonlinearity; that is, it lies entirely within the shaded sector of figure 7.2, where m_i is the sector slope. (For roundoff quantization $m_i = 2$, and for sign-magnitude truncation or overflow nonlinearities $m_i = 1$.) The result derived in [81] states the following: Given k_0 nonlinearities as described, and a $W(z)$ that is finite for $|z| = 1$, zero-input limit cycles of period N are absent if

$$\text{Re} \left\{ W(e^{j2\pi h/N}) - \text{diag} \left(\frac{1}{m_i} \right) \right\} < 0 \tag{7.2}$$

for $h = 0, 1, \ldots,$ integer$[N/2]$, where $\text{diag}(1/m_i)$ refers to the diagonal matrix whose diagonal elements all equal $1/m_i$. Furthermore, if the nonlinearities are also time invariant, with a symmetric nondecreasing characteristic, then limit cycles of period N are absent if the real part of

$$\left\{ I_{k_0} + \text{diag} \left[\sum_{l=1}^{N-1} (\alpha_{li}(1 - z_h^l) + \beta_{li}(1 + z_h^l)) \right] \right\} W(z_h) - \text{diag} \left(\frac{1}{m_i} \right) \tag{7.3}$$

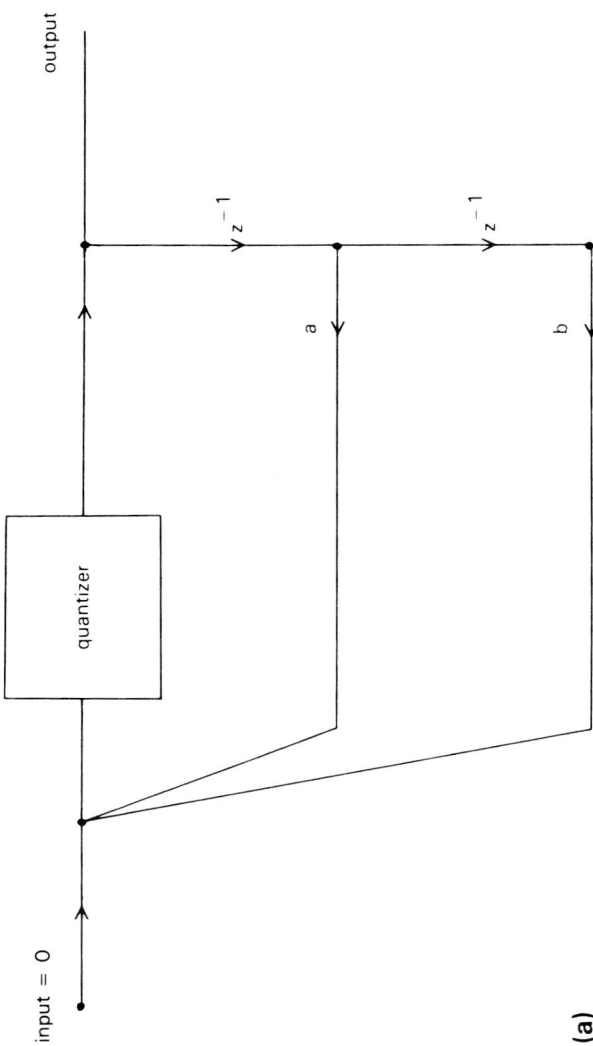

(a)

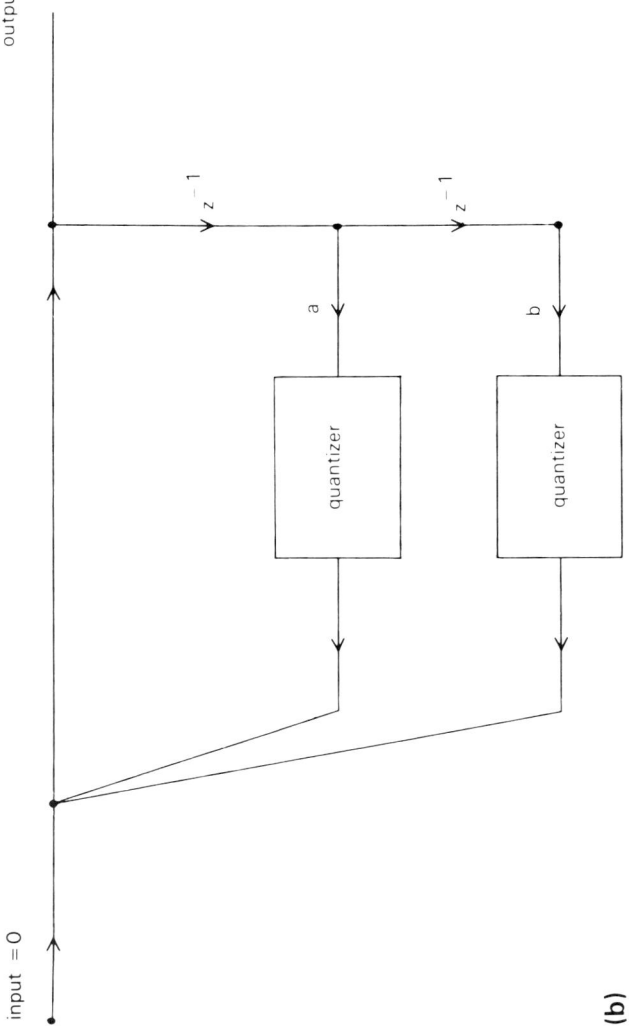

(b)

Figure **7.3** Direct form II (no zeros), 1 and 2 quantizers: (a) 1 quantizer (after double-precision adder); (b) 2 quantizers (before single-precision adder).

is negative definite (< 0), for all α_{li} and β_{li} greater than or equal to zero, where z_h is defined to be $e^{j2\pi h/N}$.

Equation (7.3) is more difficult to apply than (7.2) since linear programming techniques must be used to take advantage of the α and β parameters. However, (7.3) is a more useful condition, since it may prove nonexistence when (7.2) does not. [Note that for $\alpha = \beta = 0$, conditions (7.2) and (7.3) are identical.] If we apply (7.2) and (7.3) to the two-pole direct form II sections of figure 7.3, for both roundoff and sign-magnitude truncation nonlinearities, we can show the advantage in using sign-magnitude truncation over roundoff; the range of possible a and b values for which a quantizer limit cycle cannot occur is much greater under sign-magnitude truncation quantization [22,62].

From the computational point of view, it is unfortunate that both conditions (7.2) and (7.3) require repeated testing (once per N), not to mention the task of proving negative definiteness. We can simplify the application of (7.2) and (7.3) somewhat by expressing these conditions differently. Šiljak [82] has found an efficient technique for proving the positive realness of a function $G(z)$, which he has extended to the matrix $G(z)$ case. [A real rational function $G(z)$ is strictly circle positive real if it has no poles outside the unit circle and the real part of $G(z)$ is strictly positive on and outside the unit circle.] Thus, for one nonlinearity ($k_0 = 1$), we could replace the repeated evaluation of (7.2) with a test for the positive realness of $\{(1/m_i) - W(z)\}$. Still, this procedure is not terribly simple, especially in the matrix case, which would have to be used in considering multiple nonlinearities.

The second group of limit cycle nonexistence results for digital filters involves the norm of the transition matrix of a one-level state-space structure [83–87]. These results have been applied to both the quantizer and the overflow limit cycle. Suppose we have a one-level state-space digital filter structure:

$$v(k + 1) = f\{Av(k)\} + By(k),$$
$$u(k) = Cv(k),$$

(7.4)

where $v(k)$ are the states of the structure and f represents all the nonlinear operations of the compensator. Note that the type of nonlinearity implied by (7.4) can act *only* on the values $Av(k)$. Thus quantization must occur *after* addition, implying double-precision adders, and similarly for the overflow nonlinearity. For two's complement overflow (see section 7.2), this requirement presents no difficulty; if we define $Q(\cdot)$ to represent the two's complement nonlinearity, the following relation is true [83]:

$$Q(\eta_1 + \eta_2 + \eta_3) = Q(\eta_1 + Q(\eta_2 + \eta_3)), \tag{7.5}$$

where η_i is the result of a multiplication. The same cannot be said of the saturation overflow characteristic, and one or two extra adder bits (most significant bits) are required to accumulate the sum if more than two numbers are involved, before applying saturation. In the remainder of this work, we shall assume that all overflow characteristics are implemented so as to satisfy the requirement implicit in (7.4).

Let us consider the zero-input case $[y(k) = 0]$ for (7.4). For quantizer and overflow nonlinearities, we can show that

$$\|f(v)\|_2 \leqslant \gamma \|v\|_2 \qquad \text{for all } v, \tag{7.6}$$

where $\|v\|_2$ refers to the euclidean norm $(v'v)^{1/2}$, and should not be confused with the l_2 norm of $v(k)$ described in chapter 5. For sign-magnitude truncation and all common overflow characteristics the parameter γ would be 1, while for roundoff γ would be 2. (This is similar but not identical to the sector nonlinearity concept of figure 7.2.)

If we define the matrix norm of A as follows [84]:

$$\|A\|_2 = \max_{v \neq 0} \left\{ \frac{\|Av\|_2}{\|v\|_2} \right\}; \tag{7.7}$$

then we can write

$$\|Av\|_2 \leqslant \|A\|_2 \|v\|_2. \tag{7.8}$$

Combining (7.4), (7.6), and (7.8) gives the following:

$$\|v(k + 1)\|_2 \leqslant \gamma \|A\|_2 \|v\|_2. \tag{7.9}$$

Thus we can ensure the nonexistence of zero-input limit cycles by the condition

$$\gamma \|A\|_2 \leqslant 1 \tag{7.10}$$

since this implies a continuously decreasing state norm [84]. Mills, Mullis, and Roberts [83] have expressed this result in a different manner for the more general case of $\|v\| = (v'Dv)^{1/2}$, where D is a positive definite diagonal matrix, and the case of an overflow nonlinearity ($\gamma = 1$): Overflow (and hence sign-magnitude truncation with double-precision adders) nonlinearities will not cause zero-input limit cycles if and only if $\{D - A'DA\}$ is positive definite. (This result is based on Lyapunov theory.)

Proceeding from these results, it is natural to consider structures for which the norm of A is small (and of course less than 1). It can be shown that a minimum norm filter is one for which

$$\|A\|_2 = \max_i \{|\lambda_i|\}. \tag{7.11}$$

This quantity is always less than 1 for (stable) digital filters; thus such filter structures have no zero-input overflow oscillations and no zero-input quantizer oscillations under sign-magnitude truncation with double-precision addition.

Barnes [85] discusses minimum norm filters composed from minimum norm sections of arbitrary order. However, we shall restrict attention to the more useful case of second-order sections. In fact, a minimum norm second-order section is identical to the Rader and Gold coupled form section mentioned in chapter 6. The matrix A for a coupled form section with poles at $\sigma \pm j\omega$ would appear as follows [86]:

$$A = \begin{bmatrix} \sigma & \omega \\ -\omega & \sigma \end{bmatrix} \tag{7.12}$$

(see figure 7.4). The lack of zero-input limit cycles under overflow and sign-magnitude truncation for this structure will not be affected by scaling [83].

For roundoff quantization, these norm-based results cannot be used to prove the nonexistence of limit cycles for the minimum norm structure unless the maximum filter eigenvalue is less than one half. In fact, Jackson has shown that roundoff limit cycles *will* occur for the coupled-form structure [87]. Fam and Barnes [86] have introduced a method for taking a filter structure whose norm $\|A\|_2$ is greater than one half and computing an (infinite-precision) equivalent structure whose norm is less than one half. This technique combines recursive and nonrecursive filter sections but greatly increases the number of multipliers and delays over the original structure.

It should be mentioned here that these results tie directly into those reported for wave digital filters. Fettweis and Meerkötter [49] have shown through the use of a state norm called *pseudopower* that zero-

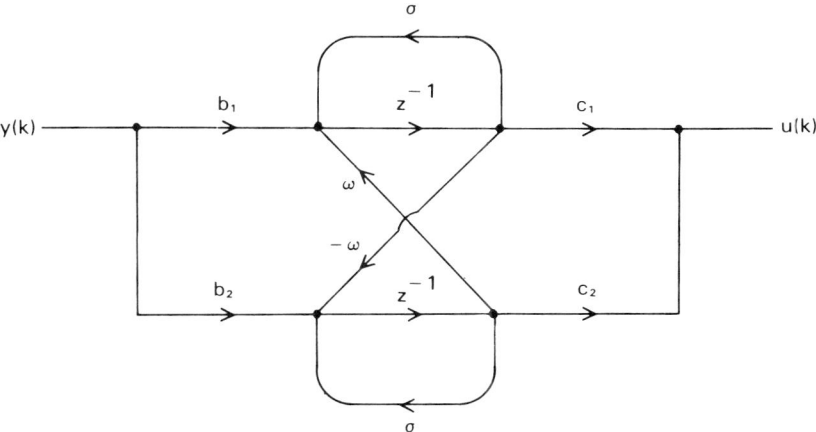

Figure **7.4** Coupled form second-order section.

input overflow limit cycles and quantizer limit cycles will not occur in wave digital filters using sign-magnitude truncation quantization and any overflow characteristic.

7.1.2 Limit Cycle Amplitude Bounds

One common method for dealing with quantizer limit cycles is to bound their amplitude and then to choose a wordlength long enough to make this bound small. Many methods exist for formulating amplitude bounds for the effects of quantization, which of course must include limit cycle effects. A good review of these methods, many of which have been presented in the context of sampled-data control systems, can be found in [62] and [88]. In the results pertaining to digital filters [88–90], the direct form II second-order section is usually considered, or specifically the recursive portion of this section. Recall that only the nonlinearities in the recursive portion of a structure can give rise to limit cycles.

The following is one of the more general approaches to limit cycle amplitude bounding. This approach involves the use of Lyapunov theory and is considered for digital filters in [88] and for sampled-data control systems in [11] and [12]. Consider a system with the following state equation:

$$x(k + 1) = Ax(k) + Bu(k), \tag{7.13}$$

where x is the state vector and u the input vector. Following the development of Parker and Hess [88], the system (7.13) is bounded-input bounded-output stable if the (zero-input) system

$$x(k + 1) = Ax(k) \tag{7.14}$$

is asymptotically stable in the large. If so, a Lyapunov function $x'Px$ exists where P is the symmetric positive-definite solution to the equation

$$P = A'PA + C \tag{7.15}$$

for any symmetric positive-definite matrix C. Thus if the input to the system (7.13) is upper bounded by some constant κ, then an upper bound on the norm of the state vector x can be derived [11, 12]. This bound, which again will include *all* the effects of quantization, is fairly complex to compute; is a function of A, B, P, and the eigenvalues of the C and P matrices; and will be directly proportional to κ.

The procedure outlined above can be easily applied to digital filters [88] with one or more precedence levels. For the roundoff nonlinearity, we know that every roundoff quantization error is bounded by $\Delta/2$ (Δ for sign-magnitude truncation). We can simply define these quantizer errors as inputs to the filter system and then compute an upper bound on the filter state norm that is proportional to Δ. The difficulty that arises in using this bound is in selecting a Lyapunov function, or, equivalently, in selecting C. Consequently, this bound can be quite loose, especially for certain combinations of filter parameters [88].

Other methods of computing limit cycle amplitude bounds either are even less tight than the Lyapunov-based bound ([9, 10]), are not easily extendible to the control system setting (such as the effective value method of Jackson [89]), or are even more difficult to compute (such as the matrix method of Parker and Hess [88]).

7.1.3 Random-Rounding Techniques For Limit Cycle Quenching

The previous two sections have described two different ways for dealing with limit cycles. The first involved using structures for which limit cycles could be proved not to exist. The second involved the use of sufficient signal bits to bound the limit cycle amplitude to a negligible level. A third method exists—actively preventing limit cycles that would otherwise occur. The idea behind this procedure is that limit cycles (which represent a correlated quantizer error effect) can be *broken up*, or *decorrelated*, by introducing some randomness into the quantization procedure. This procedure results in the replacement of a periodic limit cycle by an aperiodic sequence of reduced power [91]. Justification for this method can be found in Kieburtz [80], who reported limit cycle breakup as the level of a random input signal was raised. Further intuition for the technique can be presumed from the success enjoyed by *dither* techniques for the stabilization of unstable nonlinear systems [92, 93].

Specific results concerning the use of randomized quantization methods exist only for the case of the direct form II second-order section. The first approach involves randomly switching between roundoff quantization and sign-magnitude truncation. By utilizing roundoff *most* of the time, its low-noise advantages can still be maintained, while the occasional use of sign-magnitude truncation will give us the reduced number of limit cycles common to this type of quantizer. Kieburtz, Lawrence, and Mina [91] outline this method and present specific examples of its use. Unfortunately, such a technique cannot be guaranteed to eliminate all DC (constant value) and half-rate limit cycles (limit cycles with a two-sample period). However, Lawrence and Mina [94] do describe some additional constraints that can be added to prevent such limit cycles.

Büttner [95] has taken a different approach to implementing random quantization. In his approach, a random signal is *injected* at one node in the direct form II structure to break up any possible limit cycle. One obvious difference with this approach is that with no input to the filter, there will still be a noise output. (In a control system, already driven by noise, this additional noise would probably be insignificant.) Specifically, Büttner describes two possible approaches. First, in the direct form II section with only one quantizer (as in figure 7.3a), simply replace the least significant bit of the quantized sum with a *random* bit. This procedure produces four times the output noise power as compared to rounding, since the error introduced can be anywhere between $\pm\Delta$, but has the advantage of eliminating all possible limit cycles. The second approach introduces a random least significant bit in one of the products input to the double-precision adder. Although this generates approximately half as much noise as the first method, it will not prevent the occurrence of limit cycles unless the input to the second-order section is aperiodic and nonconstant. Büttner thus recommends using a cascade of second-order

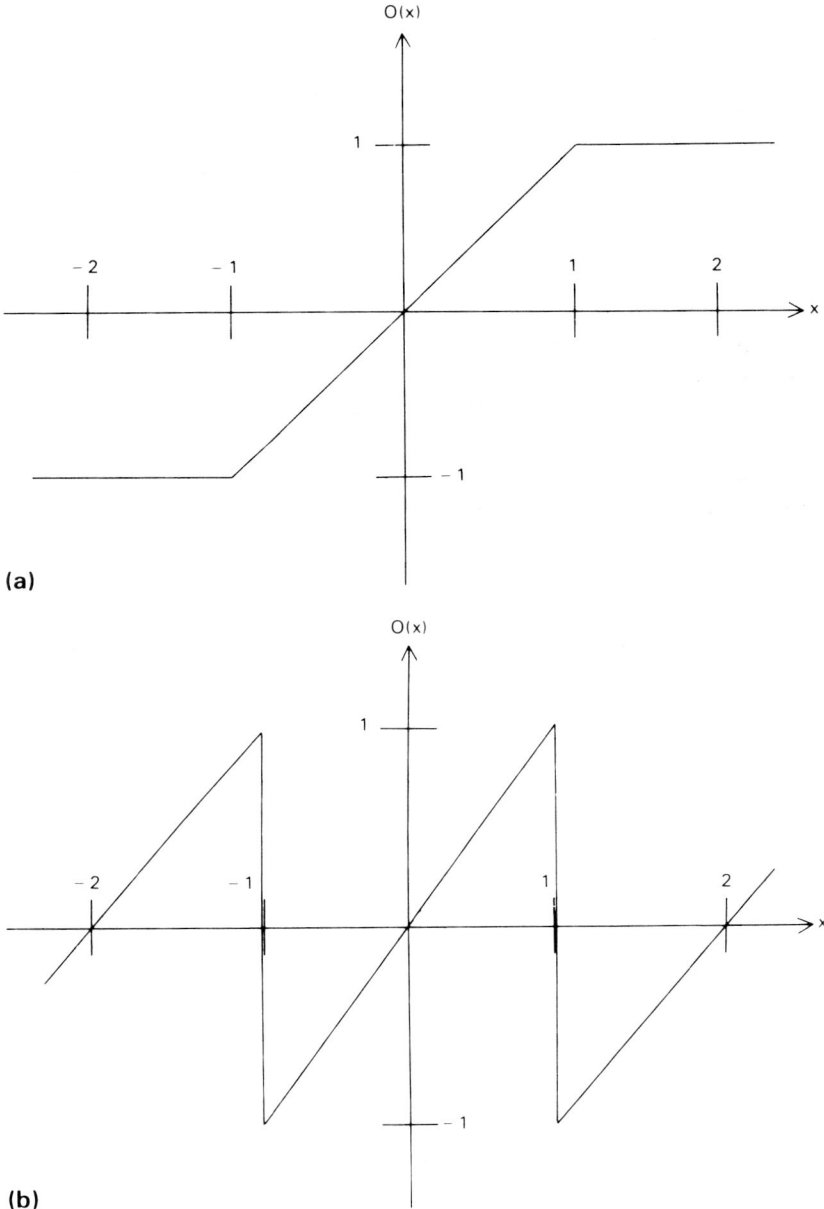

(a)

(b)

Figure **7.5** Common overflow characteristics: (a) saturation; (b) two's complement.

sections, with the first approach used to suppress all limit cycles in the first section, and the second *lower-noise* approach used in all remaining sections. Since the input to these sections must contain the random output component generated by the first section, the second method will be sufficient to suppress all limit cycles in these sections. Examples were presented comparing this random rounding approach to the use of sign-magnitude truncation to eliminate limit cycles and also to the use of roundoff quantization with longer wordlengths to reduce limit cycle amplitude. Again, all these results were generated only for structures composed of direct form II second-order sections.

7.2 Overflow Limit Cycles

In this section we shall examine the results specific to overflow limit cycles. Overflow limit cycles are particularly important because they have maximal amplitude—thus, of course, bounding techniques do not apply. In general, there are two overflow characteristics of particular interest, saturation (figure 7.5a) and two's complement (figure 7.5b). A two's complement overflow characteristic is the natural overflow characteristic resulting when using two's complement addition. No additional hardware is necessary to realize this overflow nonlinearity. The saturation overflow nonlinearity, which does require some hardware, is less prone to causing overflow limit cycles than the two's complement characteristic.

Two separate issues concerning overflow have been discussed in the digital signal processing literature: the prevention of zero-input overflow limit cycles and forced-response stability. Stability of the forced response means that the filter must *recover* from an overflow, that is, return asymptotically to the state values that would have occured if no overflow nonlinearity had been present.

General results concerning zero-input overflow limit cycles can be inferred from the discussion in section 7.1.1 on the frequency-domain criterion of Classen, Mecklenbräuker, and Peek for the saturation non-linearity (using $m_i = 1$). Using the norm-based method of Barnes and Fam, or Mills, Mullis, and Roberts, we can generate nonexistence results that would apply to *any* common overflow characteristic.

More specific results exist for the second-order direct form II section shown in figure 7.6 and for structures composed of such sections. Willson

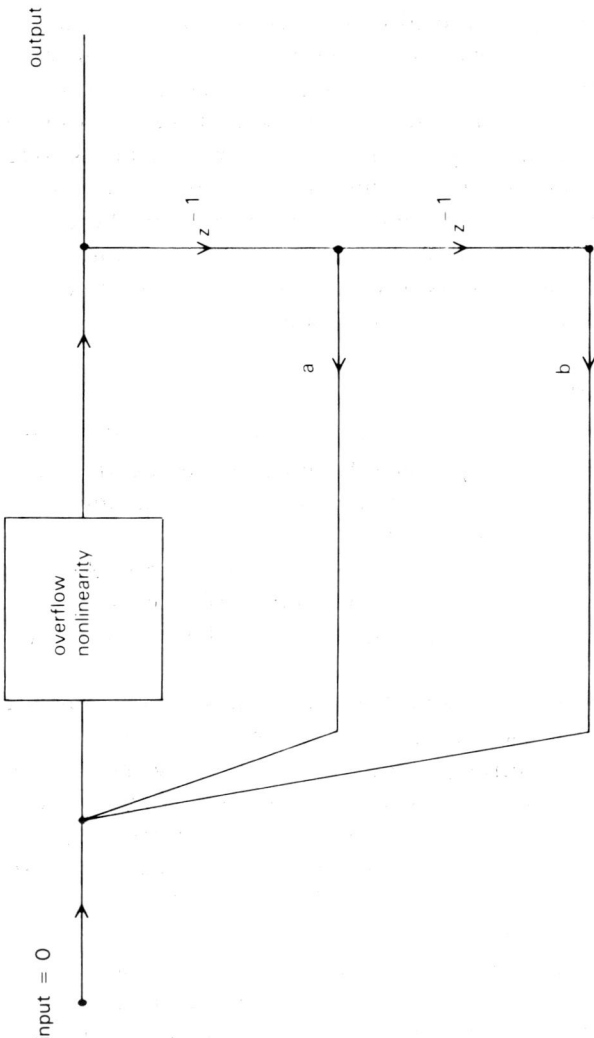

Figure **7.6** Direct form II with overflow nonlinearity.

[96] and Ebert, Mazo, and Taylor [97] have found regions in the (a, b) parameter plane where overflow limit cycles will not occur with two's complement overflow and have shown that *no* limit cycle can occur when using the saturation overflow characteristic for any (stable) values (a,b). In general, the saturation characteristic is to be preferred over the two's complement characteristic for preventing or reducing overflow limit cycles. However, it does require extra hardware components to implement the saturation overflow characteristic. Thus we would test the general conditions in section 7.1.1 to see whether the use of two's complement overflow could cause oscillations. The use of the saturation overflow characteristic, with its additional hardware, would be advised whenever the general criteria of section 7.1.1 did *not* succeed in guaranteeing the absence of limit cycles for the two's complement characteristic.

Recovery from overflow can be determined by the following general result also derived by Classen, Mecklenbräuker, and Peek [98]: If a system has no zero-input overflow limit cycles for all time-varying nonlinearities satisfying

$$-m_i \leqslant \frac{O(x,k)}{x} \leqslant 1 \qquad \text{for } x \neq 0 \text{ and } m_i > 0 \text{ for all } k, \tag{7.16}$$

where $O(\cdot)$ is the overflow nonlinearity (This condition could possibly be tested using the general criteria described in section 7.1.1), then the forced response will be stable for all overflow nonlinearities satisfying (see the shaded portion of figure 7.7)

$$\begin{aligned} 1 + m_i - m_i x < O(x) \leqslant 1 \qquad &\text{for } x > 1, \\ -1 - m_i - m_i x > O(x) \geqslant -1 \qquad &\text{for } x < -1. \end{aligned} \tag{7.17}$$

This result means that a system with no zero-input overflow limit cycles for *all* overflow characteristics satisfying (7.16) for $m_i = 1$ (such as the wave digital filter) will be forced-response stable for characteristics satisfying (7.17). Saturation satisfies (7.17), but two's complement overflow does not. Again, this result demonstrates the general advantage of saturation over two's complement overflow as far as limit cycles are concerned.

Beyond the general result of (7.16) and (7.17), there also exist specific results concerning forced-response stability for the direct form II second-order section of figure 7.6 [99, 100].

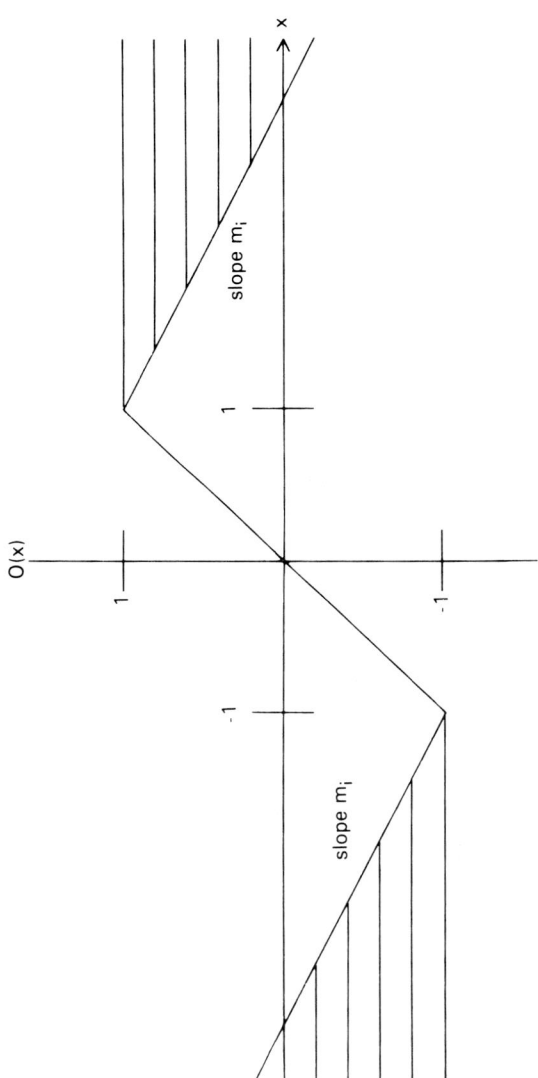

Figure 7.7 Forced-response stable overflow characteristic.

7.3 Digital Feedback Compensator Limit Cycles

In this section we shall consider the limit cycle issue as it relates to digital feedback compensators. Several important observations can be made. First, any digital control system with an open-loop unstable plant *must* exhibit some kind of zero-input quantizer limit cycle. Recall that the plant output is sampled, digitized, and quantized at the compensator input. This means that any output magnitude below the smallest quantization level is effectively ignored by the compensator. If we ignore the plant and measurement noise for now, the output of an unstable plant will tend to increase in magnitude until it reaches the lowest quantization level; then some control action can occur to drive it back toward zero. However, the process will repeat. The net result is a form of low-amplitude limit cycle in the output of the system. Such a limit cycle will occur no matter what the particular transfer functions of the (open-loop unstable) plant and compensator are, although their parameters will certainly affect the amplitude and frequency of the limit cycle. A proper choice of A/D wordlength would keep this amplitude at the system noise level, so that it could essentially be ignored. One other implication of the presence of this limit cycle is that *no* general digital filtering limit cycle nonexistence result can succeed in proving limit cycle nonexistence for digital control systems with unstable open-loop plants. Furthermore, even systems with open-loop plants that have poles at $s = 0$ can exhibit low-amplitude limit cycles if *any* offset, or bias, exists in the output of the D/A converter.

One of the key points relating to compensator limit cycles is the overall effect of the closed loop on the limit cycle behavior of the compensator. For example, consider the digital compensator as a stand-alone digital network. Any limit cycles that this open-loop compensator may exhibit are strictly dependent on the nonlinearities in the recursive sections of the compensator. However, when the compensator is embedded in the feedback loop, *all* the nonlinearities are part of a recursive portion of the control system, and thus they are all involved in determining limit cycle behavior. Consequently, compensator limit cycles that would occur for the open-loop situation will be altered when the loop is closed. By the same reasoning, even if the open-loop compensator would not exhibit limit cycles, the overall feedback system of plant and compensator together *may* exhibit limit cycles. As an example, consider the simple control system in figure 7.8. Any finite impulse response (FIR) open-loop

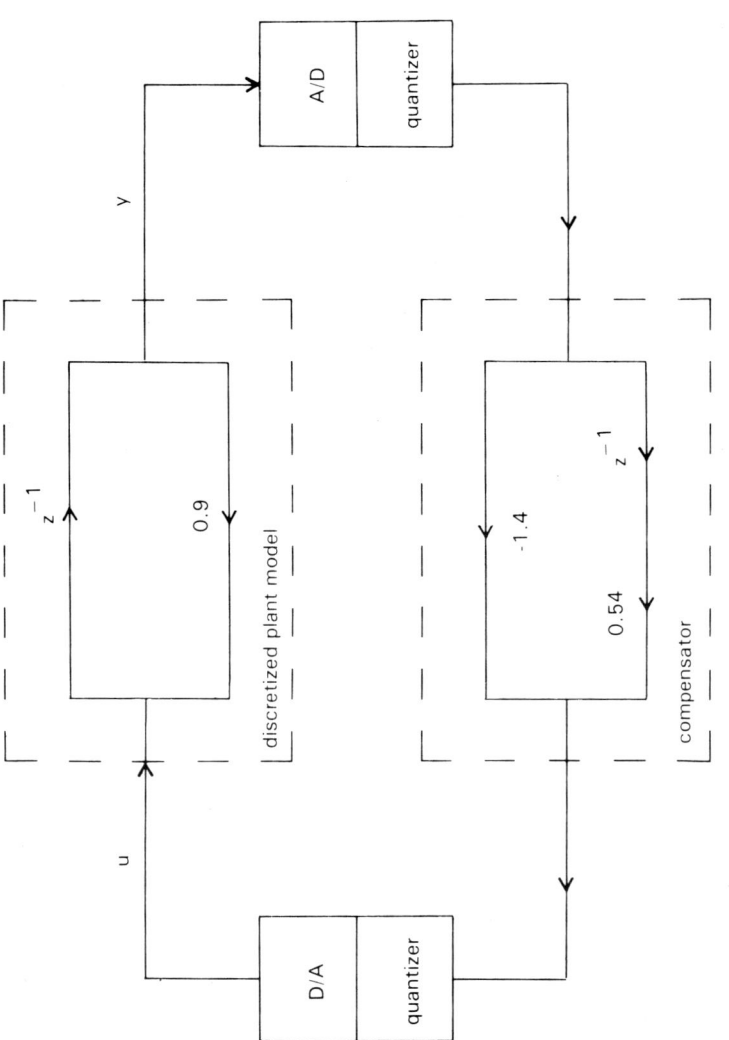

Figure 7.8 Control system with finite impulse response (FIR) compensator.

compensator or filter is nonrecursive. Therefore it can have no limit cycles. However, when we embed such a filter in a closed-loop stable control system as in figure 7.8, limit cycles may occur. For the example above, let us measure signal amplitude in units of Δ, the quantization step size defined in chapter 5. With either roundoff or sign-magnitude truncation quantization, the output y can exhibit the following half-rate limit cycle:

$$+10, \ -10, \ +10, \ -10, \ \ldots .$$

A related limit cycle result specific to feedback systems has been reported by Fettweis and Meerkötter [101]. Motivated by the presence of digital filters in looped telephone systems, they have derived the following: For a finite impulse response or wave digital filter embedded in a feedback loop, no quantizer limit cycles can occur if sign-magnitude truncation is used for all quantization operations including the A/D *and*

$$\max_{|z|=1} |H_1(z)| \max_{\omega} |H_2(j\omega)| < 1, \tag{7.18}$$

where $H_1(z)$ is the transfer function of the digital network embedded in the loop and $H_2(j\omega)$ is the transfer function of the open-loop plant. This result is quite similar to the small-loop-gain theorem known to control theorists [62]. As with the digital filtering results, the above condition points out the advantage of sign-magnitude truncation over roundoff quantization as far as limit cycles are concerned. Unfortunately, for control systems in general, the condition (7.18) is *very* restrictive in terms of the types of plants one could consider. Certainly any system whose plant had an integrator pole or even a strong resonance would not satisfy (7.18). However, this is the only real result in the literature for quantizer limit cycles in feedback systems.

Another important observation is that the techniques for dealing with limit cycles in digital filters do not tend to work for control compensators. As previously shown, *none* of the nonexistence techniques can be extended to consider open-loop unstable plants. Now let us consider control systems whose plants have integrator poles. As a simple example, consider a double-integrator plant:

$$\dot{x}[t] = \begin{bmatrix} 0 & 1 \\ 0 & 0 \end{bmatrix} x[t] + \begin{bmatrix} 0 \\ 1 \end{bmatrix} u[t],$$

$$y[t] = \begin{bmatrix} 1 & 0 \end{bmatrix} x[t]. \tag{7.19}$$

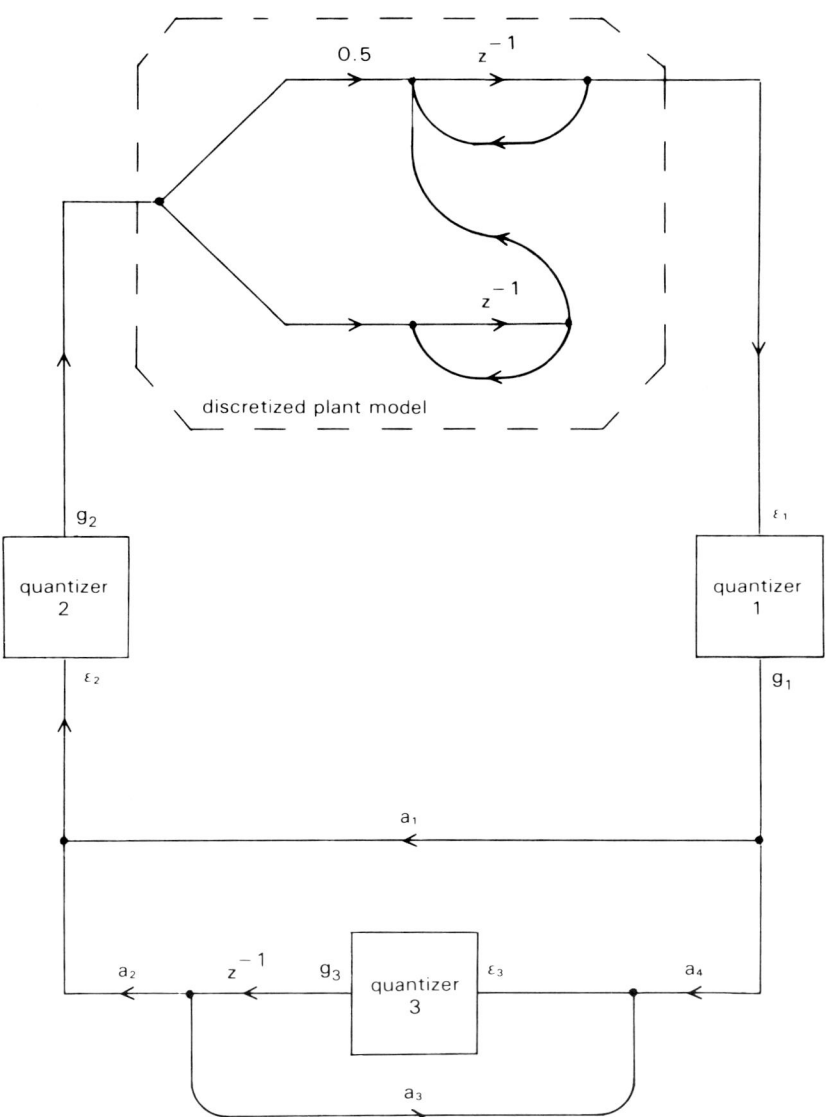

Figure **7.9** Double-integrator control system.

If we discretize this system at a sampling rate of 1 hertz and design a first-order compensator, the three-quantizer configuration of figure 7.9 results. To apply the results of Classen, Mecklenbräuker, and Peek discussed in section 7.1.1, we must first compute the matrix $W(z)$. Defining ε and g as shown in figure 7.9 yields

$$
W(z) = \begin{bmatrix} 0 & a_1 & a_4 \\ \left\{ \dfrac{z^{-1}(0.5 + z^{-1})}{(1 - z^{-1})^2} \right\} & 0 & 0 \\ 0 & \{a_2 z^{-1}\} & \{a_3 z^{-1}\} \end{bmatrix}. \tag{7.20}
$$

Unfortunately, the $(2,1)$ entry of $W(z)$ is not finite on the entire unit circle, and thus the results of (7.2) and (7.3) cannot be applied. This will be true for any system whose plant has an integrator pole. One possible method for handling this problem would be to replace the $z = 1$ poles in the $W(z)$ matrix with poles at $z = 1 - \delta$, where $\delta > 0$. Then we could evaluate (7.2) or (7.3) in the limit as $\delta \to 0$. However, this evaluation, or the application of the positive real test of Šiljak, will be even more complex to compute. Note that if a discretized plant has *all* its poles entirely within the unit circle, then the Claasen, Mecklenbräuker, and Peek results may be used directly.

Now let us attempt to apply the general norm-based results of section 7.1.1. To account for the behavior of the entire closed-loop system the vector $v(k)$ in (7.4) would have to include both the plant *and* compensator states. By the analysis of (7.6)–(7.10), this would involve the evaluation of the norm of the closed-loop system matrix analogous to the matrix A in (7.4) and the assumption that the nonlinearity f operates on the entire vector Av. For the compensator case this would be a very restrictive assumption, since in fact the actual nonlinearity only operates on the compensator states, and not on the plant states. Furthermore, the norm-based analysis applies only to one-level structures. Also, the main advantage to the norm-based technique, namely, the derivation of minimum norm structures, cannot be directly applied to compensator structures; it would involve transforming the closed-loop system matrix. However, this matrix is highly constrained, given the control system configuration of plant and compensator. Thus it cannot be subject to arbitrary transformations.

The Lyapunov-based bound discussed in section 7.1.2 has been actually used for control applications [11,12] and could even be used for open-loop unstable plants. In the analysis of section 7.1.2, let us consider the performance of the entire closed-loop system. The vector x in (7.13) and (7.14) would have to be replaced by an augmented vector x_a, which would include all the plant *and* compensator states. Of course, in the LQG case, we are not interested in bounding the norm of x_a, but the more general performance index-related norm $\|x_a' \tilde{\Upsilon} x_a\|$, where $\tilde{\Upsilon}$ is defined in (5.35). However, since $\tilde{\Upsilon}$ is a symmetric positive-definite matrix, it can be factored into some product $\tilde{\Upsilon} = T_0' T_0$. Then we can define a new x to be $T_0 x_a$, apply the appropriate similarity transformation to A and B, and proceed as outlined in (7.13)–(7.15). The resulting bound will be just as loose as for the filtering case; the difficulty will still be in selecting the Lyapunov function.

The final point we would like to make concerns the general question of limit cycles in control systems. No LQG control system is actually zero-input in nature; there is always system noise present. According to the results of Büttner discussed in section 7.1.3, it is likely that this noise will quench autonomous oscillations if the noise level is sufficiently large. Thus limit cycle oscillations themselves may not be an issue in such control systems. However, there are other effects caused by the nonlinear quantization operations in a compensator. First, jump discontinuities may occur. In such a case, small changes in the input signal lead to large jumps in the output [16]. Furthermore, we have not even considered the effects of the correlated noise that results from the presence of quantization nonlinearities. Even if limit cycles do not occur, the presence of correlated noise in control systems can significantly deteriorate performance. Recall that LQG systems are designed with the assumption that the system noises are white. This whole area is largely unexplored for digital control systems.

7.4 Summary

In this chapter we have considered the periodic effects of the quantization and overflow nonlinearities in a digital structure. We have looked in detail at the results reported for digital filters, with an eye toward their application to digital feedback compensators. Several important observations have resulted. First, control systems with open-loop unstable plants will always exhibit some kind of low amplitude zero-input limit cycle. Second, the presence of the feedback loop around a digital compensator (through the plant) changes the nature of the limit cycles that occur as compared to the case of the same compensator in an open-loop setting; in fact, the feedback loop may cause limit cycles that would not otherwise occur. Finally, limit cycles may not be a critical issue in LQG control systems at all, since the presence of the plant and measurement noises may prevent most limit cycles from occurring.

8

The Optimization
of Structures

Techniques for the optimization of structures with respect to some scalar objective function are very important in the synthesis of compensator structures. Typically this objective function would involve either the increase in the performance index due to roundoff noise, some measure of coefficient sensitivity such as the SWL or $MSWL$, or perhaps a weighted combination of the two. In such a technique, it is important to have control over the number of multipliers and delay elements in the optimized structure, since these parameters are critical in determining the complexity of the hardware.

As shown in chapter 3, any structure can be transformed to a new (infinite-precision equivalent) structure through the use of a set of transformation matrices. In the context of the modified state-space appropriate to controllers, a scaled structure with parameters $\Psi_1, \Psi_2, \ldots, \Psi_q$ can be transformed to a new scaled structure with parameters $\mathring{\Psi}_1, \mathring{\Psi}_2, \ldots, \mathring{\Psi}_q$ by

$$\mathring{\Psi}_i = P_i \Psi_i (P_{i-1})^{-1} \qquad \text{for } i = 1, \ldots, q, \tag{8.1}$$

where the P_i for $i = 1, \ldots, q - 1$ are general nonsingular transformation matrices and

$$P_0 = \begin{bmatrix} P & 0 & 0 \\ 0 & 1 & 0 \\ 0 & 0 & 1 \end{bmatrix}, \qquad P_q = \begin{bmatrix} P & 0 \\ 0 & 1 \end{bmatrix}.$$

The unity entries in the matrices P_0 and P_q are necessary so that the actual input and output nodes of the original structure are not altered by the transformation process. One consequence of this restriction is that the output node scaling parameter ρ described in section 5.1 will be invariant under such transformations.

Once we have computed (8.1), the new structure will have to be *rescaled* so that it satisfies the same dynamic range constraints as the original untransformed structure. This overall technique (transformation plus scaling) will result in a new structure with the *same* number of delay elements as the original. However, if the matrices P_i are completely general, the number of coefficients (nonunity and nonzero entries in the matrices $\overset{\circ}{\Psi}_i$) will be very large. Thus it is necessary to constrain the P_i matrices in order to gain control over the resulting number of coefficients.

Chan [17,102] has presented such a constrained optimization technique for digital filters, using a notation appropriate for describing digital filter structures. Section 8.1 will present the steps involved in this constrained optimization technique for a general objective function, but in the context of the modified state-space representation appropriate to digital feedback compensators (see chapter 3). Section 8.2 will adapt the technique of Chan for the minimization of roundoff noise effects in compensators and apply the technique to a specific example. Section 8.3 will use the *MSWL* estimate presented in chapter 6 to adapt Chan's general technique to the minimization of coefficient rounding effects in compensators. No specific example will be presented. Finally, section 8.4 will discuss methods for selecting which entries in the original Ψ_i matrices are to be constrained (held constant) and which are to be varied, presumably becoming nonzero and nonunity. This last section represents an important extension to the work of Chan since it applies equally well to digital compensators and digital filters.

8.1 The General Constrained Optimization Technique of Chan

The optimization technique of Chan is based on the following observation [17,102] (here considered in the context of the modified state-space representation). Consider the differential equation (8.2)

$$\frac{d\Psi_i(t)}{dt} = G_i(t)\Psi_i(t) - \Psi_i(t)G_{i-1}(t) \qquad \text{for } 1 \leqslant i \leqslant q, \tag{8.2}$$

where the matrices G_i are of appropriate dimension. Any solution $\{\Psi_1(t),$ $\ldots,\Psi_q(t)\}$ at any t will represent a structure (infinite-precision) equivalent to $\{\Psi_1(0),\ldots,\Psi_q(0)\}$ if

$$G_0(t) = \begin{bmatrix} G(t) & 0 & 0 \\ 0 & 1 & 0 \\ 0 & 0 & 1 \end{bmatrix}, \qquad G_q(t) = \begin{bmatrix} G(t) & 0 \\ 0 & 1 \end{bmatrix},$$

where $G(t)$ is arbitrary. Again, the unit entries are required to preserve the input and output nodes. The solution to (8.2) has the form

$$\Psi_i(t) = P_i(t)\Psi_i(0)(P_{i-1})^{-1}(t), \tag{8.3}$$

where

$$\frac{dP_i(t)}{dt} = G_i(t)P_i(t) \qquad \text{for } 0 \leqslant i \leqslant q \tag{8.4}$$

and the initial condition $P_i(0)$ matrices are identities. Starting with an initial structure that is assumed to be scaled, the technique basically involves integrating (8.4) to obtain new transformed structures. The G_i matrices are selected to cause an overall reduction in some objective function. Constraining any particular coefficient in a Ψ_i matrix to be constant can be easily accomplished by holding its derivative in (8.2) to zero, which implies constraints on G_i and P_i.

Now let us present this procedure in detail [17,102]. Define v to be the operation that forms a vector from a matrix by stacking its columns:

$$v(\Psi_i) = \begin{bmatrix} \text{column 1} \\ \text{column 2} \\ \cdot \\ \cdot \\ \cdot \\ \text{last column} \end{bmatrix}. \tag{8.5}$$

Using this operator, define $\psi(t)$ and $g(t)$ to be vectors composed of all the elements of $\{\Psi_1(t),\ldots,\Psi_q(t)\}$ and $\{G(t),G_1(t),\ldots,G_q(t)\}$:

$$
\psi(t) = \begin{bmatrix} v(\Psi_1(t)) \\ \cdot \\ \cdot \\ \cdot \\ v(\Psi_q(t)) \end{bmatrix}, \qquad g(t) = \begin{bmatrix} v(G_1(t)) \\ \cdot \\ \cdot \\ \cdot \\ v(G_q(t)) \end{bmatrix}. \tag{8.6}
$$

We can now express $d\psi(t)/dt$ as a linear function of $\psi(t)$ and $g(t)$ using (8.2) and (8.6):

$$
\frac{d\psi(t)}{dt} = F(t)g(t), \tag{8.7}
$$

where the large matrix $F(t)$ is a function of the elements of $\psi(t)$. If we wish to hold the ith component of $\psi(t)$ fixed, then we must simply set the ith component of $d\psi(t)/dt$ to zero. Thus the vector dot product of the ith row of $F(t)$ and the vector $g(t)$ equals zero. If several components of $\psi(t)$ are constrained, then let us stack up all the corresponding rows of $F(t)$ to form a matrix $R_0(t)$. Since the matrix product of $R_0(t)$ and $g(t)$ is a zero vector, we can say that $g(t)$ lies in the null space of $R_0(t)$. Thus during the optimization procedure, the vector $g(t)$ must be constrained to lie in this null space, which is a function of the elements of $\psi(t)$. Chan points out that a nontrivial $g(t)$ satisfying this constraint condition will exist if the number of ψ entries held constant is less than the dimension of $g(t)$.

The next step in the optimization procedure is to express the derivative of the objective function $f(t)$ in terms of $g(t)$. Using the chain rule and (8.4) yields

$$
\frac{df}{dt} = \sum_{i=1}^{q} \frac{df}{dP_i} \frac{dP_i}{dt}
$$

$$
= \sum_{i=1}^{q} \frac{df}{dP_i} G_i(t) P_i(t). \tag{8.8}
$$

Now, using $g(t)$ as defined in (8.6), we can introduce the gradient vector ξ and rewrite (8.8):

$$
\frac{df}{dt} = \xi'(t)g(t). \tag{8.9}
$$

We would like to select the vector $g(t)$ in the negative $\xi(t)$ direction, so that df/dt will be as negative as possible. However, keep in mind that $g(t)$ must also satisfy the null space constraint described. Thus if we choose $g(t)$ to be a unit magnitude vector indicating the direction in which the optimization should proceed while satisfying the constraint, then

$$g(t) = \frac{-\xi_R(t)}{\|\xi_R(t)\|}, \qquad (8.10)$$

where $\xi_R(t)$ is the projection of $\xi(t)$ onto the null space of R_0. As explained in Chan [17], $\xi_R(t)$ can be found by computing

$$\xi_R(t) = X(X'X)^{-1}X'\xi(t), \qquad (8.11)$$

where X is a matrix formed from a set of column vectors that constitutes a basis for the null space of R_0.

In order to create an algorithm that will implement the optimization procedure described, we must divide the continuous parameter t (call it "time") into discrete steps of length h. Thus the optimization algorithm will involve a series of computations that produces a new transformed structure at time $t + h$ from the transformed structure at time t. This process can be repeated until the value $f(t + h)$ of the objective function for the new structure is as small as we like, or until no further significant improvement seems likely.

So for a given structure at time t, we can perform all the computations involved in (8.6)–(8.11). The resulting vector $g(t)$ is used to update the transformation matrices by integrating (8.4). Chan uses the simple Euler integration formula to form a tentative $\hat{P}_i$ for the next time instant $t + h$:

$$\hat{P}_i(t + h) = P_i(t) + hG_i(t)P_i(t) \qquad \text{for } 0 \leqslant i \leqslant q, \qquad (8.12)$$

where h is the integration step size. The reason that this choice is only tentative is that the new structure formed with the transformations $\hat{P}_i(t + h)$ would not in general satisfy the scaling constraints of the original structure. We must include some scaling operation in order that the structure resulting from the transformations $P_i(t + h)$ also be scaled as desired. Recall from sections 5.1 and 5.2 that l_2 scaling involves the diagonal transformation matrices S_i whose elements are the reciprocal square roots of the diagonal elements of a set of matrices K_i. In fact, the matrices $\hat{K}_i(t + h)$ for the new (tentative) transformed structure can be related [17] to the matrices $K_i(0)$ of the original scaled structure by

$$\hat{K}_i(t + h) = \hat{P}_i(t + h)K_i(0)\hat{P}_i'(t + h) \qquad \text{for } 1 \leqslant i \leqslant q. \tag{8.13}$$

Note that the diagonal elements of $K_i(0)$ are all unity since we have assumed our original structure to be scaled. Using (8.13), we can describe the *scaling* transformations (5.14) that would have to be applied to the structure resulting from the transformations $\hat{P}_i(t + h)$ in order to scale it. In particular, the jth diagonal element of $\hat{S}_i(t + h)$ would be the reciprocal square root of the jth diagonal element of $\hat{K}_i(t + h)$. The diagonal transformation matrices $\hat{S}_i(t + h)$ can be combined with the tentative transformation matrices $\hat{P}_i(t + h)$ to form the *scaled* transformation matrices $P_i(t + h)$:

$$P_i(t + h) = \hat{S}_i(t + h)\hat{P}_i(t + h). \tag{8.14}$$

Thus the structure formed by transforming with the matrices P_i will have corresponding K_i matrices whose diagonal elements are all unity.

Using the transformations in (8.14), we can compute the new (transformed and scaled) modified state-space matrices $\Psi_i(t + h)$ with (8.3). Note that the Ψ_i matrices of the new structure are always computed by applying the updated transformations of (8.14) to the Ψ_i matrices of the original structure. In other words, the structure is *not* formed by updating the Ψ_i matrices of the previous time step. This method was used to keep the effects of numerical inaccuracy to a minimum. Even with the method currently in use, we must consider the fact that the Euler integration of (8.12) is only an approximation to (8.4). Thus after computing the new Ψ_i matrices, we must check that the constrained entries in each matrix have not changed; that is, we must check to determine whether the errors in the constrained entries are less than some preset tolerance. If these errors were too large, then one approach would be to halve the step size h and repeat the procedure starting with the computation of the tentative transformation updates $\hat{P}_i$ in (8.12). If in fact the errors are small enough, then we should reevaluate the objective function $f(t + h)$. If the resulting value is not smaller than at time t, then we should again use the approach of reducing the step size h and repeating the computations starting with the same updates in (8.12). If the objective function did turn out to be smaller than the value at time t, then the optimization procedure can continue for the next time $t + 2h$, starting with the original formation of the vector $\psi(t)$ in (8.6).

The overall algorithm can be summarized as follows:

1. Initialize the procedure with Ψ_1, Ψ_2, ..., Ψ_q as $\Psi_1(0)$, $\Psi_2(0)$, ..., $\Psi_q(0)$ and compute $K_i(0)$ as described in chapter 5. The matrices $K_i(0)$ should be identities if we start with a scaled structure. Evaluate the objective function and set all $P_i(0)$ to be identity matrices. Initialize h to 1.

2. Determine the matrix F and the constraint submatrix R_0 as defined in (8.5)–(8.7).

3. Find a set of basis (column) vectors χ_i for the null space of R_0 and form them into the matrix X:

$$X = [\chi_1 \ \chi_2 \ \chi_3 \ \cdots]. \tag{8.15}$$

4. Express the derivative of the objective function as a function of $g(t)$; that is, find $\xi(t)$ as defined in (8.9). Find its projection onto the range space of X using (8.11).

5. Evaluate $g(t)$ using (8.10).

6. Compute a tentative set of matrices $\hat{P}_i(t + h)$ by Euler integration (8.12) and evaluate the corresponding $\hat{K}_i(t + h)$ matrices in (8.13).

7. Scale the $\hat{P}_i(t + h)$ matrices using (8.14) and evaluate the new (scaled) modified state-space matrices $\Psi_i(t + h)$.

8. Check for errors in the constrained coefficients of $\Psi_i(t)$. If any, halve h and return to step 6.

9. Recompute the objective function f. If it has increased, halve h and return to step 6. Otherwise, return to step 2 unless no further improvement is desired.

8.2 The Minimization of Roundoff Noise Effects in Compensators

Chan [17,102] applied the general procedure outlined in section 8.1 to the constrained optimization of filter structures for minimum output roundoff noise variance. In this section we shall adapt this technique to the constrained optimization of *compensator* structures for minimum roundoff noise effects. In particular, the increase in the performance index J due to roundoff quantization noise will be minimized. In fact, part of this adaptation can also be used to generalize the technique of Chan for digital filters.

To apply the general technique described in section 8.1 we must specify an objective function f and also express the derivative of $f(t)$ in (8.9) as a function of $g(t)$, or in other words, compute $\xi(t)$. Chan has used an approach similar to that described in section 5.5 to form an objective function. Thus the output noise variance was expressed as a function of the matrices K_i, W_i, and Λ_i, which were discussed in section 5.5 for one-level structures. Recall that the these matrices can be found by solving two Lyapunov equations of the same order as the number of unit delays in the filter structure. Thus Chan essentially extended the roundoff noise expression derived by Mullis and Roberts and Hwang to apply to multiple-level filter structures. Chan was then able to define an objective function and derive an expression for its derivative as necessary in (8.9).

In this section we shall adapt Chan's roundoff noise expression to the digital compensator case. Specifically, we shall use the context of the modified state-space representation, account for the performance of the entire closed-loop system, and also specify the objective function to reflect the increase in the performance index J. Thus the expression derived in section 5.5 will be extended to the case of multiple-level compensator structures [see (5.45)–(5.50)]. We shall also show that the expression derived by Chan for the derivative of f applies almost unchanged to the compensator case.

Equation (5.51) can be extended to include multiple precedence levels as follows. Excluding A/D noise, we can rewrite equation (5.34) as (again using tildes to distinguish the quantities of the scaled system)

$$\tilde{Z} = \tilde{A}\tilde{Z}\tilde{A}' + \frac{\Delta_r^2}{12}\begin{bmatrix} 0 & 0 \\ 0 & \tilde{\Omega} \end{bmatrix}, \tag{8.16}$$

where $\tilde{A}$ is defined in (5.33) and

$$\tilde{\Omega} = \Lambda_q + \tilde{\Psi}_q\Lambda_{q-1}\tilde{\Psi}'_q + \tilde{\Psi}_q\tilde{\Psi}_{q-1}\Lambda_{q-2}\tilde{\Psi}'_{q-1}\tilde{\Psi}'_q$$
$$+ \cdots + \tilde{\Psi}_q\cdots\tilde{\Psi}_2\Lambda_1\tilde{\Psi}'_2\cdots\tilde{\Psi}'_q.$$

Recall that Λ_i is a diagonal matrix whose jth diagonal element represents the number of roundoff noise sources associated with the jth row of $\tilde{\Psi}_i$, Δ_r is the quantization step size of the quantizers in the structure, A contains the parameters of the closed-loop system, and Z is the steady-state covariance matrix of the plant and compensator states. Also note that the parameter k_{ad} will have no effect on the optimization procedure described below, or on the procedure to be described in section 8.3. Thus it can be set to 1 as stated in section 5.1.

If we replace $\tilde{Z}$ with TZT^{-1} as in (5.47), where T is the scaling transformation matrix that relates the original unscaled system of plant and compensator to the scaled system,

$$T = \begin{bmatrix} I_n & 0 \\ 0 & S_q \end{bmatrix}, \tag{8.17}$$

then (8.16) can be rewritten

$$Z = AZA' + \frac{\Delta_r^2}{12} \begin{bmatrix} 0 & 0 \\ 0 & \hat{\Omega} \end{bmatrix}, \tag{8.18}$$

where A is given in (5.46) and

$$\hat{\Omega} = \Lambda_q S_q^{-2} + \Psi_q \Lambda_{q-1} S_{q-1}^{-2} \Psi_q' + \Psi_q \Psi_{q-1} \Lambda_{q-2} S_{q-2}^{-2} \Psi_{q-1}' \Psi_q'$$
$$+ \cdots + \Psi_q \cdots \Psi_2 \Lambda_1 S_1^{-2} \Psi_2' \cdots \Psi_q'.$$

The expression for dJ, the increase in the performance index due to roundoff noise, is given in (5.47) and repeated for reference:

$$dJ = \text{trace}\{\Upsilon Z\}. \tag{8.19}$$

Using the adjoint Lyapunov equation as described in appendix B, and as applied in (5.48)–(5.51), we can express dJ as

$$dJ = \frac{\Delta_r^2}{12} \text{trace} \left\{ W \begin{bmatrix} 0 & 0 \\ 0 & \hat{\Omega} \end{bmatrix} \right\}, \tag{8.20}$$

where W is given in (5.49). Defining the lower right-hand $(n + 1) \times (n + 1)$ portion of W to be W_q, we can rewrite (8.20):

$$dJ = \frac{\Delta_r^2}{12} \text{trace}(\Lambda_q S_q^{-2} W_q$$
$$+ \Lambda_{q-1} S_{q-1}^{-2} \Psi_q' W_q \Psi_q + \Lambda_{q-2} S_{q-2}^{-2} \Psi_{q-1}' \Psi_q' W_q \Psi_q \Psi_{q-1}$$
$$+ \cdots + \Lambda_1 S_1^{-2} \Psi_2' \cdots \Psi_q' W_q \Psi_q \cdots \Psi_2). \tag{8.21}$$

Once we have reached this point, the remainder of the development is very similar to the development of Chan [17]. As Chan has done, we can now define the matrices W_i. Using a recursive definition,

$$W_i = \Psi_{i+1}' W_{i+1} \Psi_{i+1} \qquad \text{for } i = 1, \cdots, q - 1. \tag{8.22}$$

These matrices are called *noise gain* matrices by Chan, since their diagonal elements reflect the gain from each roundoff noise source variance to the variance of the filter output. For the compensator case, they will represent the gain from each roundoff noise source variance to the increase dJ in the performance index. By means of (8.22), equation (8.21) can be further simplified:

$$dJ = \frac{\Delta_r^2}{12} \sum_{i=1}^q \text{trace}\{\Lambda_i S_i W_i\} \tag{8.23}$$

or equivalently,

$$dJ = \frac{\Delta_r^2}{12} \sum_{i=1}^q \left\{ \sum_j [\Lambda_i]_{jj} [S_i^{-2}]_{jj} [W_i]_{jj} \right\}. \tag{8.24}$$

Thus only the diagonal elements of W_i appear in (8.24). Since the diagonal elements of K_i equal the diagonal elements of S_i^{-2}, we can replace the middle term in (8.24) with $[K_i]_{jj}$. Since the scale factor $(\Delta_r^2)/12$ will not affect the minimization process in any way, we can formulate the following objective criterion for the effects of roundoff noise:

$$f = \sum_{i=1}^q \left\{ \sum_j [\Lambda_i]_{jj} [K_i]_{jj} [W_i]_{jj} \right\}. \tag{8.25}$$

We explicitly show the $[K_i]_{jj}$ term in f, although it is equal to the identity matrix for any already scaled structure because it is important when deriving derivative expressions.

Now we must turn to the task of expressing the derivative of f as a function of $g(t)$. (We shall also drop the tilde superscript and assume scaling, as in section 8.1.) Chan [17] has shown that the digital filtering K_i and W_i matrices have the following derivatives:

$$\frac{dK_i(t)}{dt} = G_i(t)K_i(t) + K_i(t)G_i'(t),$$

$$\frac{dW_i(t)}{dt} = -G_i'(t)W_i(t) - W_i(t)G_i(t). \tag{8.26}$$

These will apply equally well to the compensator case with its analogous K_i and W_i matrices. Using (8.24)–(8.26) and following the method used in Chan, we can write the derivative of the objective function f:

$$\frac{df}{dt} = \sum_{i=1}^q \left\{ \sum_j [\Lambda_i]_{jj} \left(\left(\frac{d}{dt} [K_i]_{jj} \right) [W_i]_{jj} + [K_i]_{jj} \left(\frac{d}{dt} [W_i]_{jj} \right) \right) \right\}. \tag{8.27}$$

After substituting for the derivatives in (8.27) with the expressions in (8.26) and some manipulation as in Chan [17], we arrive at the following compact expression:

$$\frac{df}{dt} = \sum_{i=1}^{q} \text{trace}(M_i'(t)G_i(t)), \tag{8.28}$$

where

$$[M_i(t)]_{jk} = 2\{[K_i(t)]_{jk}[W_i(t)]_{jj}[\Lambda_i]_{jj} - [\Lambda_i]_{kk}[W_i(t)]_{jk}\}. \tag{8.29}$$

The quantity ξ needed in the optimization procedure can easily be obtained from (8.28) and (8.29).

As in chapter 5, the K_i and W_i matrices in (8.29) are not defined exactly as those defined for digital filters. Other than this, there are two external differences between our expression in (8.29) and the original expression derived by Chan. First, the lack of the factor $[K_i]_{kk}$ in the second term of (8.29) is due to the fact that all the diagonal elements of K_i are unity; recall that it was assumed that the original structure was scaled. This is largely a procedural difference between Chan's derivation and ours—in terms of the optimization, it makes no difference. The second difference in this expression is the presence of the Λ_i term. Recall that the jth diagonal entry of Λ_i represents the exact number of roundoff noise sources associated with the jth row of the scaled matrix Ψ_i. During the optimization procedure, any unconstrained unity or zero entries in these Ψ matrices will in general become nonzero and nonunity. Thus these new sources must also be included in the Λ_i matrices at the beginning of the optimization procedure. Inclusion of the Λ_i terms allows us to consider all possible structures. The assumption made by Chan, that the Λ_i matrices can be taken to be identities or proportional to identities, can often be an oversimplification, especially for structures with multiple precedence levels and few coefficients. In such cases, it will result in a suboptimal structure. However, the inclusion of the Λ_i terms can easily be incorporated into the digital filtering optimization techniques of Chan. This is one example where the techniques developed in this work can be applied to digital filter analysis.

With the optimization procedure derived by Chan and the correct inital conditions, a structure identical to the minimum roundoff noise filter structure derived in closed form by Mullis and Roberts can be found. Similarly, using the adapted optimization procedure described and the correct initial conditions, we can also duplicate the minimum roundoff noise compensator structure derived in section 5.5. To achieve this result for compensator structures, we must allow all the coefficients (except the next-to-last column of Ψ_1) of a one-level initally scaled structure to vary. Thus all the diagonal entries of the matrix Λ_1 must be set to $n + 1$. Similarly, by allowing only 2×2 diagonal blocks of coefficients, plus the last row and column (input and output coefficients) of Ψ_1 to vary, we can optimize and produce a block optimal structure. In fact this procedure was used to generate the (one-level) block optimal F8 compensator structure studied in chapter 5 and 6. (For this example and the following, an A/D scale factor of 0.25 was accounted for as mentioned in section 5.6.)

The optimization procedure was also applied to the (scaled) two-level parallel F8 compensator structure composed of direct form II sections and designated as (c) in chapters 5 and 6. Its modified state space (before optimization) is

$$\Psi_2 = \begin{bmatrix} 1 & 0 & 0 & 0 & 0 & 0 \\ 0 & 1 & 0 & 0 & 0 & 0 \\ 0 & 0 & 1 & 0 & 0 & 0 \\ 0 & 0 & 0 & 1 & 0 & 0 \\ 0 & 0 & 0 & 0 & 1 & 0 \\ 0 & 0 & 0 & 0 & 0 & 1 \\ c & c & c & c & c & c \end{bmatrix},$$

$$\Psi_1 = \begin{bmatrix} 0 & 1 & 0 & 0 & 0 & 0 & 0 & 0 \\ c & c & 0 & 0 & 0 & 0 & 0 & c \\ 0 & 0 & 0 & 1 & 0 & 0 & 0 & 0 \\ 0 & 0 & c & c & 0 & 0 & 0 & c \\ 0 & 0 & 0 & 0 & 0 & 1 & 0 & 0 \\ 0 & 0 & 0 & 0 & c & c & 0 & c \end{bmatrix},$$

$$(8.30)$$

where each entry c represents a coefficient. Two extra coefficients were added by allowing two other entries in Ψ_1 to vary, the (5,5) and (5,6) entries. Thus there were 17 coefficients total in the optimized structure. For this example, the matrices Λ_1 and Λ_2 were

$$\Lambda_1 = \text{diagonal}[0 \ 3 \ 0 \ 3 \ 2 \ 3],$$
$$\Lambda_2 = \text{diagonal}[0 \ 0 \ 0 \ 0 \ 0 \ 6]. \tag{8.31}$$

Before optimization, the scaled coefficient values ranged from 10.48 to 0.073. Table 8.1 shows the range of coefficient values and the resulting number of signal bits necessary to hold the value of dJ due to roundoff to 5% (as in chapter 5) after each iteration of the optimization process.

Without including the 2 extra coefficients, which altered Λ_1 and increased the apparent required wordlength of the initial structure to 10.46, the number of bits needed (see table 5.2) was 10.18. Thus the true improvement resulting from the optimization was 2.12 bits. This is quite impressive, since it was attained basically in only two iterations and is quite close to the block optimal value of 7.88 bits, which requires 25 coefficients. In fact, it is identical to the performance of the 17-coefficient parallel structure (b). We note that iterations 3, 4, 5, and 7 involved halving the integration step size due to increases in dJ over the current least value and that the value after iteration 9 was actually lower than the value after iteration 2, but not appreciably. In the digital filter examples treated by Chan in [17, 102], typically only 5 to 8 iterations were required to achieve the full benefits of the optimization procedure. The block optimal compensator structure computed via this optimization procedure took 5 iterations to reach the approximate minimum wordlength.

One byproduct of the optimization procedure for the example of table 8.1 was a reduction in the maximum coefficient value. Instead of needing 4 integer bits to represent the largest coefficient (see chapter 6), the optimized structure required only 1, again an impressive savings in

Table **8.1** Roundoff noise optimization results

Iteration number	Number of bits	Coefficient range
0	10.46	10.48–0.073
1	8.745	3.20–0.108
2	8.057	1.46–0.108
3	8.2	1.46–0.108
4	8.086	1.46–0.108
5	8.06	1.46–0.108
6	8.056	1.46–0.108
7	8.057	1.46–0.108
8	8.056	1.46–0.108
9	8.055	1.46–0.108

wordlength. Intuitively, this savings may exist for any increase in the number of coefficients in a structure. This point needs more investigation.

8.3 The Minimization of Coefficient Wordlength in Compensators

In this section we shall develop an objective function for the minimization of coefficient rounding effects on the performance index J. We shall use the $MSWL$ expression as presented in chapter 6. The optimization could just as well be carried out for the SWL estimate, but the $MSWL$ is simpler to compute and potentially tightly related to the more accurate SWL. This objective function is *quite* different from the one developed in Chan [17], since it is based on J, and hence involves second-order sensitivities. Again, as with the SWL and $MSWL$ derivations, this development will be useful in digital filtering for filters that are designed by optimizing some scalar differentiable criterion.

Instead of minimizing the actual $MSWL$ value, we shall use an approach similar to that of the previous section; we shall minimize the expected value of dJ due to coefficient rounding. Of course, for the analysis of finite wordlength coefficient effects, this expected value is taken over an ensemble of structures—it is not a time average as in the roundoff noise case. A review of the results of chapter 6 indicates that $E(dJ)$ can be written

$$E(dJ) = \frac{\Delta^2}{24} \sum_{m=1}^{N} \left(\frac{\partial^2 J}{\partial c_m^2} \bigg|_x \right), \tag{8.32}$$

where N is the number of quantized coefficients in the scaled structure. Recall that all entries in the precedence level matrices of a structure whose ideal values cannot be represented exactly with a certain finite number of bits are subject to coefficient quantization. Thus all zero, integer, and simple power-of-$(1/2)$ entries would not be considered in this analysis.

To form a simpler objective function, the constant factor can be dropped from (8.32) as in the previous section:

$$f = \sum_{m=1}^{N} \left(\frac{\partial^2 J}{\partial c_m^2} \bigg|_\infty \right), \tag{8.33}$$

where

$$\frac{\partial^2 J}{\partial c_m^2} = \text{trace} \left(\tilde{\Upsilon} \frac{\partial^2 \tilde{Z}}{\partial c_m^2} \right), \tag{8.34}$$

$$\tilde{Z} = \tilde{A}\tilde{Z}\tilde{A}' + \begin{bmatrix} \Theta_1 & 0 \\ 0 & \{\Psi_{12}\Theta_2\tilde{\Psi}'_{12}\} \end{bmatrix}, \tag{8.35}$$

and $\tilde{\Upsilon}$ contains the performance index weighting matrices as shown in (5.35). The tilde will again be used to distinguish the parameters of the scaled system. We can also write the expressions (8.34) and (8.35) for the compensator before scaling:

$$\frac{\partial^2 J}{\partial a_m^2} = \text{trace}\left(\Upsilon \cdot \frac{\partial^2 Z}{\partial a_m^2}\right), \tag{8.36}$$

$$Z = AZA' + \begin{bmatrix} \Theta_1 & 0 \\ 0 & \{\Psi_{12}\Theta_2\Psi'_{12}\} \end{bmatrix}, \tag{8.37}$$

where a_m represents a coefficient of the unscaled structure. As with the approach for roundoff noise minimization discussed in section 8.2, we would like to express f as a function of the unscaled parameters and the scaling matrices S_i. This is necessary for the computation of the derivative of f; even though the original structure selected for the optimization will be scaled and its S_i matrices will thus be identities, they do affect the derivative of f.

The terms $(\partial^2 J)/(\partial c_m^2)$ can be related to the terms $(\partial^2 J)/(\partial a_m^2)$ as follows. Since $\tilde{\Psi}_i = S_i\Psi_i(S_{i-1})^{-1}$, a scaled coefficient c_m at index (j,k) in the matrix $\tilde{\Psi}_i$ can be related to its unscaled counterpart by

$$[\tilde{\Psi}_i]_{jk} = [S_i]_{jj}[\Psi_i]_{jk}[(S_{i-1})^{-1}]_{kk}. \tag{8.38}$$

Since c_m is thus a multiple of a_m, we can write

$$\frac{\partial^2 J}{\partial c_m^2} = \frac{\partial^2 J}{\partial a_m^2}\left(\frac{[S_{i-1}]_{kk}}{[S_i]_{jj}}\right)^2. \tag{8.39}$$

We can express this relation compactly for all the coefficients in level i as

$$Y_2(\tilde{\Psi}_i) = S_i^{-2}Y_2(\Psi_i)S_{i-1}^2, \tag{8.40}$$

where $Y_2(M)$ is a matrix function whose dimensions match those of its argument matrix M and whose (j,k)th element is $\partial^2 J/\partial[M]_{jk}^2$ only if the (j,k)th location in M corresponds to a quantized coefficient in the scaled structure and is zero otherwise.

For the optimization procedure described in section 8.1, we shall need to determine the relation of the second-order sensitivities of J with respect to the coefficients of the *transformed* structure to the second-order sensitivities of J with respect to the untransformed coefficients. However, the general transformation matrices P_i in (8.1) and (8.2) are not diagonal, so the simple expression in (8.40) cannot be used. In fact, for the coefficient c_m in the ith level of the transformed structure, the term $(\partial^2 J)/(\partial c_m^2)$ will now be related to all the second partials of J with respect to the entries in Ψ_i that correspond to quantized coefficients in the transformed structure, including the *mixed* second partials. To demonstrate this, the following matrix chain rule can be applied [103]: If x and y are scalars and M a matrix, then

$$\frac{\partial y}{\partial x} = \text{trace}\left(\frac{\partial y}{\partial M}\frac{\partial M'}{\partial x}\right). \tag{8.41}$$

For this expression, the derivative of a scalar with respect to a matrix M is defined to be the matrix whose (j,k)th element is the derivative of the scalar with respect the (j,k)th element of M.

Now let us designate the precedence level matrices associated with the transformed structure by a superscript circle, as in (8.1). By applying the matrix chain rule with J as y, the coefficient at index (j,k) in $\overset{\circ}{\Psi}_i$ as x, and Ψ_i as M, we get

$$\frac{\partial J}{\partial [\overset{\circ}{\Psi}_i]_{jk}} = \text{trace}\left(\frac{\partial J}{\partial \Psi_i}\frac{\partial \Psi_i'}{\partial [\overset{\circ}{\Psi}_i]_{jk}}\right). \tag{8.42}$$

Recall from (8.1) that the relation between the transformed and untransformed precedence level matrices can be written $\overset{\circ}{\Psi}_i = P_i\Psi_i(P_{i-1})^{-1}$. Thus the second term in the trace of (8.42) can be written

$$\frac{\partial \Psi_i'}{\partial [\overset{\circ}{\Psi}_i]_{jk}} = P_{i-1}' E_{kj}(P_i')^{-1}. \tag{8.43}$$

Note that equation (8.42) seems to imply that the derivative $\partial J/\partial [\overset{\circ}{\Psi}_i]_{jk}$ is a function of the matrix $\partial J/\partial \Psi_i$, which involves derivatives with respect to *all* the entries of Ψ_i, not just the few coefficient entries. This would imply a tremendous computational load, especially when second derivatives were considered—for a seventh-order compensator with two precedence levels and seven intermediate nodes, we would have to compute mixed second partials with respect to all the entries in each level, or $49^2 + 49^2$ second derivatives. This number would be independent of the number of actual quantized coefficients in the structure. Even though this computation need be performed only once at the start of the optimization process, it would involve far too much computation time.

Fortunately, it is not necessary to compute all the derivatives above. In fact, since the matrices P_i are constrained not to vary certain fixed entries in the original Ψ_i matrices, the matrix in equation (8.43) will have a special property—it will have zero entries in exactly those locations that will eliminate the dependence of (8.42) on derivatives with respect to Ψ_i entries that are not in the same locations as the quantized coefficients of the transformed structure. In other words, the derivatives of J with respect to the quantized coefficients in the transformed structure will be functions only of the derivatives of J with respect to the Ψ_i entries that are in the same exact locations as those coefficients. To reflect this fact, the term $\partial J/\partial \Psi_i$ in (8.42) should be replaced by $Y(\Psi_i)$, where $Y(M)$ is a matrix function whose dimensions match those of its argument matrix M and whose (j,k)th element is $\partial J/\partial [M]_{jk}$ only if the (j,k)th location in M corresponds to a quantized coefficient (nonzero, nonunity, and so forth) in the transformed structure and is zero otherwise. Note that this definition of $Y(M)$ is analogous to the definition of $Y_2(M)$ in (8.42). Thus we can rewrite (8.42):

$$\frac{\partial J}{\partial [\mathring{\Psi}_i]_{jk}} = \text{trace}(Y(\Psi_i) P'_{i-1} E_{kj}(P'_i)^{-1}). \tag{8.44}$$

To relate the second derivatives of J with respect to the coefficients of the transformed and untransformed structures, let us take the derivative of (8.44):

$$\frac{\partial^2 J}{\partial [\mathring{\Psi}_i]_{jk}^2} = \text{trace}\left\{ P'_{i-1} E_{kj}(P'_i)^{-1} \frac{\partial}{\partial [\mathring{\Psi}_i]_{jk}}(Y(\Psi_i)) \right\}. \tag{8.45}$$

Inside the trace expression, the matrix chain rule (8.41) can be applied to each nonzero element of the derivative of $Y(\Psi_i)$. For example, if the (r,s) entry of Ψ_i is also a quantized coefficient, then

$$\frac{\partial}{\partial [\mathring{\Psi}_i]_{jk}}([Y(\Psi_i)]_{rs}) = \text{trace}\left\{ P'_{i-1} E_{kj}(P'_i)^{-1} \frac{\partial}{\partial \Psi_i}\left(\frac{\partial J}{\partial [\Psi_i]_{rs}} \right) \right\}. \tag{8.46}$$

We shall define this trace to be $[B_i]_{rs}$. Interchanging the order of differentiation, and applying the same reasoning that eliminated the extra derivative terms in (8.42), we can express (8.46) as follows:

$$[B_i]_{rs} = \text{trace} \left\{ P'_{i-1} E_{kj} (P'_i)^{-1} \frac{\partial}{\partial [\Psi_i]_{rs}} (Y(\Psi_i)) \right\}. \tag{8.47}$$

Note the presence of the mixed second partial derivatives of J in (8.47). Let us define the matrix B_i to have nonzero entries $[B_i]_{rs}$ as described in (8.47), where (r,s) is the location of a quantized coefficient in the transformed structure, and zero entries otherwise. With this definition, (8.45) can be rewritten

$$\frac{\partial^2 J}{\partial [\Psi_i]^2_{jk}} = \text{trace} \{ P'_{i-1} E_{kj} (P'_i)^{-1} B_i \}. \tag{8.48}$$

Thus with (8.47) and (8.48), we have now fully described the relation between the second partial derivatives of J with respect to the coefficients of the transformed structure and the mixed second partials of J with respect to the corresponding coefficients in the untransformed structure.

Since the transformation matrices P_i are general, they will almost certainly not produce a scaled structure even if the structure *before* transformation was scaled. However, we can include scaling in this formulation by applying the results derived in (8.38)–(8.40) to the transformed structure. Thus the complete expression for the objective function $f(t)$ will be

$$f = \sum_{i=1}^{q} S_i^{-2} Y_2(\mathring{\Psi}_i) S_{i-1}^2, \tag{8.49}$$

where

$$[Y_2(\mathring{\Psi}_i)]_{jk} = \text{trace} \{ P'_{i-1} E_{kj} (P'_i)^{-1} B_i \} \tag{8.50}$$

and $\mathring{\Psi}_i$ represents the coefficient parameters *after* transformation and *before* scaling, and B_i is a function of all the mixed second partials of J with respect to the entries in Ψ_i that correspond to coefficients in the transformed scaled structure. Given that the transformed scaled structure will have N coefficients, the advantage to the above formulation of f is that the N^2 mixed second-order coefficient sensitivities need only be computed once, at the start of the optimization procedure. The N^2 Lyapunov equations that will have to be solved for these sensitivities [similar to those in equation (6.22)] cannot be simplified through any application of the adjoint Lyapunov operator described in appendix B. However, the solutions to these equations can be simplified in the same manner as the computations described in section 6.3 for solving (6.21). Specifically, the first step of the Lyapunov solution method can be by-passed for all but one of the N^2 equations. As before, this saves most of the total computation time involved in such solutions.

Now that an expression for $f(t)$ has been formulated, we can examine its derivative with respect to the transformation parameter t. Following the procedure of (8.8), we must first evaluate the derivatives of f with respect to the matrices P_i and then multiply the resulting ith term by the matrices $G_i P_i$ for all i. From (8.49), df/dP_i will involve the derivative of S_i^{-2}, which is a matrix composed of the diagonal elements of K_i, and the derivative of S_{i-1}^2. These can be found by applying (8.26) and a simple matrix identity for the derivative of a matrix inverse [103]. The derivative df/dP_i will also involve the derivative $dY_2(\mathring{\Psi}_i)/dP_i$. This term can be computed directly since the expression for $Y_2(\mathring{\Psi}_i)$ in (8.50) and involving B_i from (8.47) contains specific terms dependent on P_i. All the other terms in (8.47) and (8.50) are not dependent on P_i. The actual formation of $\xi(t)$ in (8.9) from the resulting derivative expressions will be quite tedious, but really is only a matter of bookkeeping. As a whole, the method described is computationally quite efficient. However, we have not tested the optimization procedure of section 8.1 in the context of the statistical wordlength-based objective function (8.49) for an actual example.

8.4 Criteria For Selecting Unconstrained Coefficients

As stated by Chan [102], one of the major open issues concerning this optimization procedure relates to the selection of the unconstrained entries in the Ψ_i matrices. For the optimization of parallel or cascade compensator structures composed of second-order sections, we have formulated some general guidelines that seem appropriate. As will be shown, these guidelines can be applied equally well to the digital filtering case.

For the optimization of roundoff noise effects, the block optimal form of Mullis and Roberts and of Hwang still tends to have too many coefficients, as compared with structures of nearly the same performance. However, it is possible to use minimum roundoff noise second-order sections combined with direct form II sections, thereby saving several coefficients. In order to select the section that should be converted to a minimum roundoff noise section, we must examine the objective function f in (8.25). Recall that f depends only on the diagonal elements of the matrices Λ_i and W_i. The matrices Λ_i reflect the number of roundoff sources that are associated with the rows of the scaled matrices Ψ_i, and the diagonal elements of the matrices W_i contain the gains from the variances of the roundoff noise sources r_i to performance index J. For a parallel direct form II structure [see (3.24) and figure 3.9], which has two levels, the diagonal elements of W_1 will be pairwise associated with the specific second-order sections. Since we know the weights $[\Lambda_i]_{jj}$, the relative diagonal values of W_i will indicate which sections in the structure contribute the most to the objective function f. The matrix W_2 for a parallel direct form II structure will not be important to this consideration, since Ψ_2 contains coefficients, and hence roundoff sources, that only affect the output node. Recall from (8.1) that this node cannot be altered by the optimization procedure.

Let us consider the example treated in section 8.2. For this structure, the parallel structure (c), the diagonal values of W_1 were as follows:

$$[W_1]_{jj} \qquad \text{for } 1 \leqslant j \leqslant 6 = \{1.71, 7.32, 0.092, 0.264, 342, 465\}. \qquad (8.51)$$

Since the diagonal values of W_1 are pairwise associated with the three second-order sections of this example, we can easily identify the third section as the trouble spot—the third pair of values $(342, 465)$ is clearly the largest, given the weights Λ_1 in (8.31). This fact justifies the specific location of the two extra coefficients chosen to be varied. In fact, if we had allowed the section to be truly a minimum roundoff noise section, it would have required three extra coefficients and not two. However, in this example the results indicate that the performance with two is quite excellent—hence one should not automatically go to a block optimal section. Certainly, this point requires further investigation.

When optimizing only a portion of a structure as discussed here, it is necessary to know the performance level that would result for the block optimal case, so that one can judge the effectiveness of using fewer coefficients. This value can be found using this same optimization procedure, but with more unconstrained Ψ_i entries (more coefficients). Note that this approach to determining which section of a structure to optimize can also be adapted to include cascade structures. We should also mention that the above guidelines will of course not be too effective if the diagonal elements of W_1 tend to be similar in magnitude.

A similar guideline may be used when minimizing coefficient word-length. As mentioned in chapter 6, by computing the $MSWL$ or SWL, we have already computed the second partials of J with respect to the coefficients in the structure. Furthermore, the SWL computation will also produce the *mixed* second partials of J. It is precisely these sensitivities that we need to produce f in (8.49). We would simply have to compute the SWL of the original structure $\{\Psi_i(0)\}$ and save the sensitivities. If any of the second-order sensitivities $(\partial^2 J)/(\partial a_m^2)$ of the original structure are particularly large as compared to any others, then the second-order section in which those coefficients reside would be a likely candidate for optimization. In particular, any zero or unity entries in the portion of the Ψ_i matrix corresponding to that section should be unconstrained in the optimization procedure. Such a section, when optimized, will have the same form of modified state space representation as a minimum roundoff noise section, but it will be optimal with respect to a different criterion.

Although the criteria presented above by no means fully answer the question of which Ψ_i entries to constrain, they do provide an important guide in situations in which performance *and* minimal numbers of co-efficient multiplies are important.

In one sense, the constraint issue is part of a larger topic: the selection of an initial structure from which to optimize. One property of the iterative constrained optimization procedure described in this chapter is that the number of precedence levels is fixed during the optimization. Therefore, optimizing a two-level structure for some objective function does not tell us whether an extra level will significantly improve performance, or whether one less level can be used without degrading performance. In general, more levels provide more degrees of freedom for the optimization, but of course this will depend on the number of constrained coefficients and their locations in the Ψ_i matrices. For now, these questions must be dealt with by trying different initial structures, with different numbers of levels. Further work is needed in this area, for both the synthesis of digital filter structures and the synthesis of digital compensator structures.

8.5 Summary

In this chapter we have adapted a procedure described by Chan for syn-thesizing optimal filter structures to the case of the digital feedback compensator. A major advantage to this procedure is that specific entries in the modified state-space matrices can be held constant to control the number of coefficients and thus also the number of multiplications in the structure.

We have described efficient computational techniques for synthesizing compensator structures that optimize roundoff quantization noise and coefficient rounding effects, both expressed as increases in the performance index J. We have presented two examples for the constrained minimization of roundoff noise effects and closed with a discussion of some guidelines for the selection of unconstrained coefficient locations. Although this last area remains largely unexplored, we believe that the observations made apply equally well to digital filter optimization.

9

Summary,
Conclusions, and
Future Efforts

In this section we shall review the issues discussed and the results developed in this work, with stress on their applicability to digital compensators, on the one hand, and digital filters, on the other.

Many elegant mathematical solutions exist for control problems. Often, the resulting compensators are directly implemented on large-scale computer systems, where speed and accuracy are assured, and cost not critical. The issues involved in the implementation of such compensators on small-scale digital systems have not received the attention they deserve. For these applications, the finite memory, relatively slow speeds, and the expense of the hardware *must* be considered in the overall design process. Fortunately, these very issues have been examined in the context of digital signal processing, and a great many useful results exist. Our approach was to *use*, *adapt*, and *extend* these results to digital feedback compensators. In several situations, however, we have extended these results to the point where they also constitute a useful extension for digital filtering applications. These extensions will also be pointed out in this summary.

The steady-state LQG control and estimation problem was selected as a basic framework for several reasons. First, this type of controller has been shown to have desirable performance properties in terms of its robustness, multivariate formulation, optimal nature, and so forth. Second, the LQG problem has received a great deal of attention in the recent literature and is being increasingly applied to real systems. Third, the LQG problem has an explicit scalar objective function, which can be adopted as a performance metric against which the degradation due to finite wordlength effects can be measured. It is not necessary to choose such a performance measure or even the LQG problem at all. However, this choice allows us to develop results in a concrete setting. Finally, using the LQG control framework makes it possible to raise all the important issues, and this can in fact be done using single-input single-output systems. As we shall argue, extensions to the multiple-input multiple-output case are straightforward, although the issue of multiple-input multiple-output structures remains largely unexplored.

Chapter 2 presented the assumptions, problem statement, and solution method involved for an LQG system and raised a key point. The calculations involved in producing the compensator output and state values require a finite amount of time t_c. This time must be accounted for in the LQG design procedure. Two implications arise: (1) the sampling period must be greater than t_c, and (2) the compensator output at a given sample time can only depend on *past* compensator state and input values. However, if $T \gg t_c$, we must not constrain the system to wait a full T seconds for its control update. It *should* only have to wait t_c seconds. Hence, we presented the LQG solution method and sample-skew concept from Kwakernaak and Sivan [1].

Once such an *ideal* compensator is designed, it must be implemented in finite-precision hardware. In chapter 3 we presented the concept of a *structure* as defined for digital filters and the notation introduced by Chan for representing such structures. The concept of an accurate notation for reflecting the arithmetic and quantization operators in a structure and the inherent precedence of these operations is critical; although all structures have the same transfer function and same performance as the ideal compensator *under infinite precision*, they will in fact all differ, in general, under finite-precision arithmetic. For control applications, two points were stressed. First, a state space is insufficient to represent all possible structures (which leads to the notation developed by Chan). In fact, it can represent only that class of structures possessing one precedence level. Second, and more important, the notation developed by Chan for filter structures is not quite suitable for representing compensator structures since the concept of a structure is slightly different in control applications. In digital filtering, the calculation time necessary to compute the next filter output from the current filter states is ignored, since it only represents series delay time. Whether the filtered data emerges 0.1 seconds after its input or 0.15 seconds is really of no concern, as long as the data rate is high enough for the application. However, this delay *must* be included in any compensator structure, since this structure is embedded in a feedback loop. If one considers this delay as part of the plant (that is, as a series delay following the compensator), then this effectively raises the dimension of the plant and of any compensator designed via the LQG approach. We can avoid this by considering the delay to be part of the compensator. Since the LQG design technique of chapter 2 has the compensator output depending only on past inputs, we can always express the compensator structure in such a way that the output node is a state node (the output of a delay branch). This will account for the time required to compute the *next* output from the current compensator states and input. Thus we have modified the notion of a structure to more accurately reflect all the delays that exist. Consequently, all compensator structures shall include the output node as a state node.

This altered notion of a structure requires a new notation, called the *modified state-space representation*. It has all the advantages of Chan's notation for digital filters and furthermore includes *all* the calculation delays that exist in the compensator. One major implication of this definition of compensator structures is that a delay-canonic structure (one that has a minimal number of delays) for an nth-order compensator has $n + 1$ unit delay elements, instead of n as in a digital filter. Thus a cascade of direct form I second-order sections, not canonic for digital filters, *is* canonic for digital compensators. In the context of this definition of structures, we presented several classes of structures and pointed out that a straightforward implementation of the ideal compensator equations (called a "simple" structure) is *not* usually a good choice for steady-state LQG compensators, since it has many more coefficients than nearly every other structure used in digital filtering. Of course, for situations in which it was not convenient to compute the parameters of any structures other than the simple structure, such as in adaptive control systems or in any system in which the appropriate Ricatti equations must be computed on line, the simple structure or a one- or two-level version of this structure (still with many coefficients) must be used.

Chapter 4 presented several digital computer architecture concepts as they related to digital filters and digital compensators. The basic idea of serialism and parallelism, the degree to which processes run sequentially or concurrently, extends without modification to digital compensators. The intuition that can be gained concerning precedence and maximally parallel architectures from the Chan notation for digital filters is identical to that gained from the modified state-space representation for digital compensators. However, the same cannot be said concerning the application of pipelining to compensators. In fact, the application of pipelining to compensators brings out another point—the interaction between the ideal design process discussed in chapter 2 and the implementation of the resulting compensator. Basically, the use of pipelining alters a structure so that the number of precedence levels in the structure is reduced, while still producing *nearly* the same transfer function. The only difference is the addition of one or more series delay units. Fewer precedence levels means a smaller minimum calculation time and a *faster possible sampling rate*. For digital filters, the extra series delay encountered is of no importance. What is significant is the potential increase in the data rate. However, for compensators this delay *must* be considered in the design process. If ignored, this delay results in extra negative phase shift and the performance of the control system may deteriorate—it may even become unstable, as demonstrated in chapter 4. To include the effects of the delay, we can now simply increase the order of the plant (with one additional state per unit delay added) and redesign the optimal LQG compensator for the higher sampling rate. The resulting higher-order compensator structure must be able to be pipelined in the same manner as was the original structure. Depending on the application, the pipelined control system with its increased sample rate can have superior performance as compared to the original, slower, nonpipelined system.

In the next three chapters, the effects of finite wordlength in digital compensators were investigated. These effects were divided into three areas: the uncorrelated effects resulting from quantization of the multiplier products and the compensator input (quantization noise, chapter 5), the correlated effects of these same quantization operations and the overflow nonlinearities in the compensator (limit cycles, chapter 7), and the effects of quantizing the infinite-precision coefficients of a structure (coefficient quantization, chapter 6).

The analysis of quantization noise includes an important subissue—scaling. Scaling is necessary to match the dynamic range of the signals in the structure to the dynamic range representable with the fixed-point words. Various types of scaling were described for digital filters, depending on the known characteristics of the compensator input signal; some are more conservative than others (because they assume less restrictions on the input), thereby resulting in higher noise levels. For digital feedback compensators, two issues were brought out. First, the common LQG set-point configuration makes use of a compensator with *two* inputs, either or both of which can have DC components. This fact would require that the most conservative type of scaling be used (l_1 scaling), and would in fact require the use of techniques for scaling multiple-input structures. However, the use of an *alternative* but equivalent set-point configuration can avoid this problem. With the alternative configuration, the compensator typically has only one input, and this input has no DC component. Thus a less conservative scaling procedure can be employed. The *stochastic* l_2 scaling method applied equalizes the probability of overflow at every node in the structure. However, this probability depends on the behavior of the *entire* closed-loop system, not the compensator alone (which could be unstable without the feedback path). Thus we have adapted this digital signal processing scaling procedure for use with digital *compensators* by including the effects of the closed loop.

Once a structure is scaled, we can compute the effect of quantization noise on some objective criterion. For digital filters, we presented the modeling associated with *roundoff* and *sign-magnitude truncation* quantization and restricted the analysis to the more tractable (and lower-noise) case of roundoff. To compute the noise power due to roundoff errors at the output of a digital filter, a Lyapunov equation of order $n + 1$ can be solved, where $n + 1$ is the number of unit delays in the filter. For digital compensators, again, the effect of roundoff errors on the performance index is a *closed-loop* phenomenon. Thus we have *adapted* the analysis method to include the entire plant and compensator system, as done for compensator scaling. In addition, for digital signal processing applications, Mullis and Roberts have derived a one-level minimum roundoff noise filter structure. It proved possible to adapt this method to produce a minimum roundoff noise compensator structure. As before, the entire closed-loop system had to be considered.

To test the roundoff effects of different structures for implementing a higher-order compensator, the F8 example was introduced. The results from a roundoff analysis of these structures brought out several points. First, as in digital filtering, the direct form II structure had poorer performance in terms of the increase in J due to roundoff noise than factored forms like the cascade or parallel structures, and also as in digital filtering, the pairing and ordering issues associated with cascade structures were significant in determining their performance. As expected, the block optimal minimum roundoff noise compensator structure performed better than any of the other structures tested. However, two points were raised that were different for digital compensators as compared to digital filters. First, the pairing issue for control compensators is further complicated by the presence of many *real* poles. Most digital filters have at most one real pole while controllers can frequently have more than one. These poles must be paired if second-order sections are to be used; the same applies to the real zeros. Thus even a parallel compensator structure brings out the pairing issue, where parallel *filter* structures have no such consideration. Secondly, the default "simple" structure for digital compensators, not used for filter structures, did perform comparatively well. However, there were two structures with many fewer coefficients that performed even better.

The effect of *coefficient* rounding on performance is basically deterministic. Given a set of coefficients, we can compute exactly the resulting performance degradation. However, in digital filtering, a statistical approach based on first-order sensitivities has been developed for estimating the coefficient wordlength required to meet some degradation level. Thus it is not necessary to evaluate the performance repeatedly until a suitable wordlength is found. We have extended this statistical approach to the LQG compensator and in so doing have raised an important point. Because the LQG design technique minimizes the performance index J with respect to the compensator parameters, all first-order sensitivities with respect to the compensator coefficients are zero. Thus second-order sensitivities are necessary to estimate the increase in J due to coefficient rounding, and of course J can only increase with such rounding. The necessity for such second-order terms will be true of *any* parameter optimization problem; for example, certain reduced-order compensator design techniques would fall into this category. In fact, if a *digital filter* is designed to minimize some differentiable scalar objective function, then a statistical wordlength estimate for this filter using this same objective function must also use second-order sensitivities. This then constitutes an extension to the results for the implementation of digital filters.

Other issues concerning coefficient wordlength were raised when the statistical methods described were applied to the F8 system. First, we have evaluated the structures according to the wordlength required to achieve a specific degradation level. As in digital filters, there was a strong correlation between the low noise and low coefficient sensitivity structures. Again, it was observed that the "simple" structure performed well, but was still outperformed by the same two structures as in the roundoff analysis. The new *SWL* statistical estimate, developed using second-order sensitivities, proved to be conservative, as is its filtering counterpart based on first-order sensitivities. However, for the five structures requiring the least bits, it was very accurate (conservative by at most 1.4 bits). The *SWL* value was much more conservative for these poorer structures: the direct form II, and the cascade and parallel structures using identical inadvisable pole pairings. *Unlike* the usual digital filter statistical estimate, a second simpler-to-compute estimate was possible, based only on the mean degradation in performance. (This value would be zero for any estimate based on first-order sensitivities.) This *MSWL* estimate was *very* tightly related to the *SWL* value, from 0.68 to 0.94 bits lower in all 10 cases, and can thus easily be used for a relative wordlength comparison between several candidate structures or in an optimization algorithm. The major advantage of these two statistical estimates over a deterministic determination of wordlength was *not* in the computation time saved, which was minimal (15–30%) for under 20 coefficients and negative (it took longer) for over 20, but in one other very important area. Since the estimates were continuous in nature and differentiable, either could be used as the scalar objective function for a structural optimization procedure. In such a procedure based on the statistical estimate, we had to compute all the (mixed) second partial derivatives of J with respect to the N coefficients — but this was required only once for the entire iterative procedure. This point was further developed in chapter 8.

  In the discussion on limit cycles in chapter 7, we reviewed the methods used in digital filtering for dealing with limit cycles. Although new results in this area were limited, four observations relating to digital compensators were brought out. First, a control system with an open-loop unstable plant, or a plant with an integrator pole, must of necessity have some sort of low-amplitude zero-input limit cycle. The system output will increase from zero until it reaches the lowest quantization level of the output A/D. Only then can control action seek to restore the system to the zero level — but then the process will repeat. This situation is unavoidable since the system is essentially *open-loop* when the magnitude of the output level is less than one A/D quantization level. Second, the global feedback loop around the compensator will change the nature of the limit cycles in the compensator and can even cause limit cycles. For example, a finite impulse response filter will not exhibit limit cycles, yet a feedback system using a finite impulse response compensator may exhibit limit cycles. Third, the techniques used in filtering for dealing with limit cycles do not often extend to compensators, especially when the plant has a pole at the origin or in the right half-plane. Finally, from the random rounding and other experimental results in the digital signal processing literature, it is not clear whether any significant limit cycles will exist in LQG systems. The noise driving the system and the measurement noise in the output will tend to quench any limit cycle that may occur. This of course will depend on the intensity of the noise. However, even though limit cycles themselves may be suppressed, other nonlinear effects such as jump discontinuities may occur. Furthermore, the quantization noise in the system is *not* white, and the very presence of correlated noise in the system may cause difficulties. There are few techniques for handling these effects, even for digital filters.

The final topic treated in this work was the iterative constrained optimization of structures. The basis for this technique lies in the work of Chan for filters. However, we can again adapt the algorithm to handle digital compensators. For minimizing roundoff effects, the adaptation was quite similar to that required to compute the closed-form block optimal one-level minimum roundoff noise structure of chapter 5. However, for minimizing coefficient rounding effects, the extension is quite different from the Chan approach, since the statistical estimate is based on *second*-order sensitivities. We demonstrated the optimization technique for roundoff noise effects for several structures, but did not test the changes required to produce a minimum coefficient wordlength structure. This effort in optimization did bring out two points that extend Chan's optimization technique for digital filters also. First, the technique presented in chapter 8 for the constrained minimization of roundoff noise was more general than Chan's. We accounted for the exact number and location of roundoff error sources in the structure; Chan uses an approximation to simplify his analysis. This change can easily be incorporated into Chan's filter structure optimization algorithm. Second, we pointed out some general approaches to selecting the portion of a compensator structure that will produce the greatest improvement when optimized. These guidelines also apply to the optimization of filter structures.

Based on all these results, there are several extensions that should be mentioned, and also several new issues that we did not address. Let us first consider some of the extensions, to both other performance criteria and other control or estimation problems.

In principle, the results developed extend to the consideration of other performance measures, such as gain margin and phase margin. In actuality, the details of the derivations and the actual equations will be quite different. For example, the statistical wordlength estimate may be dominated by first-order sensitivities. However, for the steady-state Kalman filter problem (considered at length by Sripad [13]) and its performance measure, the results would be more directly applicable. As in the LQG case, this problem has a simple minimized scalar objective function, the trace of the error covariance matrix. However, since this is not a control problem, but an estimation problem, it will have many of the characteristics of a digital filter. Thus, while a statistical wordlength procedure for the Kalman filter will require the use of second-order sensitivities (like the LQG case), the scaling and roundoff analysis procedures will not depend on any *closed-loop* system behavior (unlike the LQG case). Still, the adaptation of our results and techniques to digital Kalman filter implementations will be fairly straightforward. Of course, the Kalman filter may have to be treated as multiple output (see the discussion below on multiple-input multiple-output systems).

Our efforts can also be easily extended to certain suboptimal parameter optimization control problems. Both the optimal nature and the closed-loop aspects of the LQG problem are found in these controllers. In fact, if the same J is taken to be the performance measure, all the results apply. The equations will differ only in the fact that, in general, the compensator dimension will be smaller than the plant dimension.

One important extension of the results is to multiple-input multiple-output compensators, since there are a great many real-world systems that are multiple-input multiple-output in nature. Given some multiple-input multiple-output structure, our results apply with only a few minor changes. However, the whole question of how one designs multiple-input multiple-output structures is basically unexplored. The modified state-space notation is sufficiently flexible to cover the multiple-input multiple-output case, if we simply allow input and output vectors, instead of scalars. *Multiple-output* scaling is no problem, since the present technique already scales all the nodes. However, some modifications will be required to implement scaling procedures for *multiple-input* LQG compensators. Certainly we can still compute the variances of all the nodes of the compensator, accounting for the closed-loop nature of the control system and its driving and measurement noises. Recall that the aim of the stochastic scaling procedure was to equalize the probability of overflow at all the compensator nodes and the compensator input (plant output). However, for multiple-output plants (multiple-input compensators), there is a problem. Figure 9.1 shows a simple double-input compensator. The variances of the two system outputs y_1 and y_2 will not, in general, be the same. Thus we cannot equalize the probabilities of overflow at every node and every compensator input unless we select different word-lengths for the two input A/D converters. One possible solution is to select only one of the compensator inputs to have the same probability of overflow (after scaling) as all the nodes and to allow the remaining compensator inputs to have a lower probability of overflow. This can be accomplished by choosing the compensator input y_i with the largest variance for use in the scaling procedure of chapter 5. Instead of normalizing K_q in (5.22) and K_i in (5.24) by dividing by the variance of y, we shall use the variance of y_i. However, in equation (5.23) the symbol y must refer to the vector y, not y_i. Other than these changes, the rest of the compensator scaling procedure basically remains the same. (In the full multiple-input multiple-output scaling procedure, recall that u must also be a vector.) One other point involving scaling should be mentioned. Since each A/D unit has its own scale factor, we must also consider this scaling issue in the multiple-input sense: To preserve the overall ideal system performance, all the A/D scale factors must be the same (unless a coefficient multiply could be combined in the scale factor of an A/D). Again, the actual value will depend on the plant output whose combined variance and/or system transients are the *largest*.

The question of multiloop limit cycles does not really further complicate the limit cycle question. If any effective limit cycle analysis method is found for dealing with single-loop control systems, it should directly extend to the multiloop case.

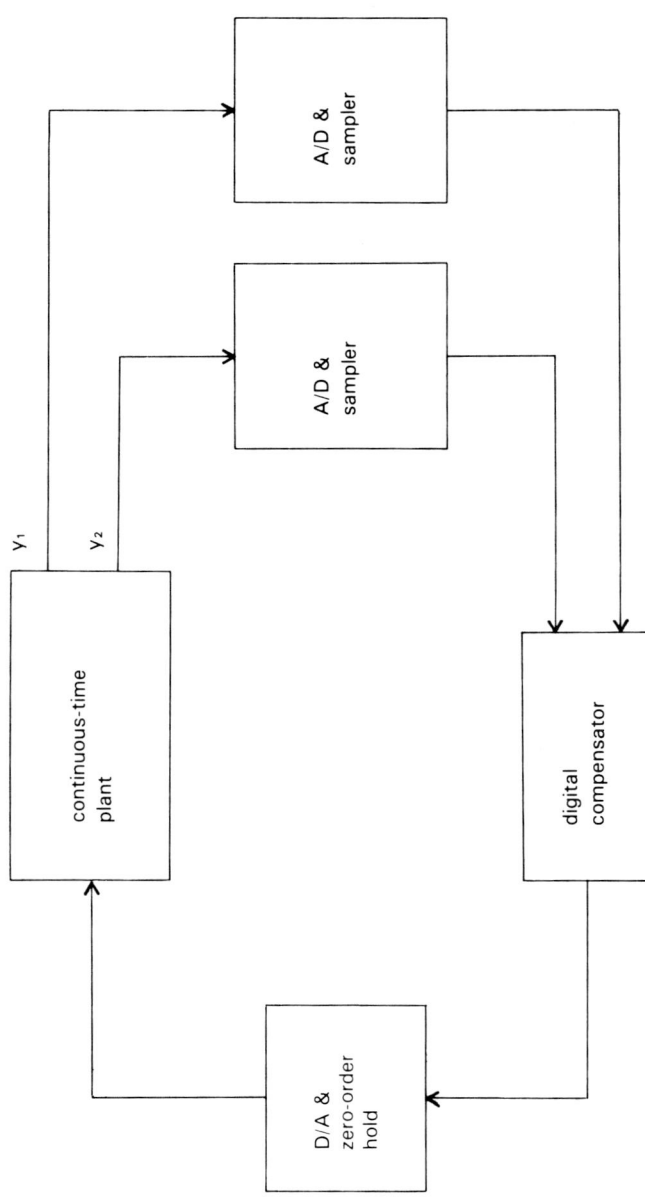

Figure **9.1** Double-input compensator control system.

As mentioned, there are several issues that we did not consider in this work. The first of these involves the nature of the LQG problem. By expressing all the desired performance characteristics of a control system in a single all-encompassing scalar function J, there can be some question as to the relevance of the connection between the minimization of J and the satisfaction of the initial performance objectives. The work of Harvey and Stein [25] and Stein [26] mentioned in chapter 2 is an important step toward solving this problem. What we can state is the following: To the extent that the index J is relevant to the desired control system performance, analyses based on increases in J will be relevant to the *relative* performance of an implementation.

Limit cycles themselves may not be an issue for LQG control systems. However, there is a middle ground between white additive quantization noise and a limit cycle oscillation. Jump phenomena and the presence of correlated noise can be very detrimental to control applications. The work of Sripad [13], Sripad and Snyder [60], and Parker and Girard [104] on the correlated nature of quantization errors should serve as a foundation for studying such effects.

Another important issue involves the constrained optimization of structures discussed in chapter 8. At one level, more work needs to be done in testing and evaluating the minimization of coefficient wordlength. However, on a more fundamental level is the question of how to select the initial structure. (Recall that the iterative optimization procedure must begin with a specific structure and then apply transformations to it.) The choice of initial structure is important because the iteration procedure cannot change the number of precedence levels in the initial structure. The question of how many precedence levels to use is very complex. It is dependent on the number of (unconstrained) coefficients desired, the speed requirements of the application, and the acceptable level of performance degradation. Furthermore, given an initial structure, we do not always know the best way to choose *which* coefficients to constrain. Such considerations are of importance in the optimization of both digital compensator structures and digital filter structures.

It is also important to rethink the issue of arithmetic type. This effort has assumed right from the start that a fixed-point numerical representation is being used. Fixed-point arithmetic implies minimal expense and minimal computation time as compared to floating-point arithmetic computation. However, as the hardware evolves, new systems of arithmetic arise that may be competitive with fixed-point. For example, a system called FOCUS [105] has been reported in the literature. The main motivation for FOCUS has been the problems encountered in control and certain other signal-processing applications. Specifically, control systems require the most accurate control signals when the system output is close to the desired level (to reduce steady-state error) and less accurate control levels when far from the desired set point. The FOCUS system of numerical representation and arithmetic combines the accuracy advantages of floating point with the hardware simplicity and higher speed of fixed point. Applications of this work on compensator implementation to the FOCUS number system may become quite useful for control systems.

Finally, as a closing point, we would like to consider an open area that may become of great interest to digital control system designers. The implicit objective of this work has been to take an ideal discrete-time compensator design technique and implement it with as little degradation as possible, subject to certain constraints. There was never any question that the design technique itself was to consider anything other than the infinite-precision case. However, Johnson [106] has pointed out a fundamental question with regard to general digital control system design: Since the compensator for a continuous-time plant is to be implemented in finite precision, probably with some kind of finite-state machine, why not directly design the compensator so that it is optimal with respect to that finite-state machine? Johnson has also observed that the problems involved in designing such "finite-state compensators" are so complex and loosely defined that it will take the combined efforts of many researchers to solve them. However, the potential benefits are great.

The purpose of this research effort has been to expose the fundamental issues involved in the digital implementation of control compensators and to use, adapt, and extend the techniques of digital signal processing in order to develop methods applicable to control. We believe that these efforts have provided a foundation for an overall methodology for the implementation of compensators.

Appendix A

F8 Data

This appendix will present the continuous-time F8 model discussed in chapters 5 and 6 and its discrete-time equivalent. The G and K matrices computed by the procedures mentioned in chapter 2 are also given. Finally, data defining all 10 candidate structures analyzed in chapters 5 and 6 and also the optimized structure discussed in chapter 8 will be presented.

These are the parameters of the sixth-order single-input single-output continuous-time F8 system, following the notation of chapter 2:

The A matrix for the continuous-time sixth-order F8 system:

$-6.696d-01$	$5.7000d-04$	$-9.010d+00$	$0.000d+00$	$-1.577d+01$	$0.00d+00$
$0.000d+00$	$-1.3457d-02$	$-1.411d+01$	$-3.220d+01$	$-4.330d-01$	$0.00d+00$
$1.000d+00$	$-1.2000d-04$	$-1.214d+00$	$0.000d+00$	$-1.394d-01$	$0.00d+00$
$1.000d+00$	$0.0000d+00$	$0.000d+00$	$0.000d+00$	$0.000d+00$	$0.00d+00$
$0.000d+00$	$0.0000d+00$	$0.000d+00$	$0.000d+00$	$-1.200d+01$	$1.20d+01$
$0.000d+00$	$0.0000d+00$	$0.000d+00$	$0.000d+00$	$0.000d+00$	$0.00d+00$

The B matrix:

$0.000d+00$
$0.000d+00$
$0.000d+00$
$0.000d+00$
$0.000d+00$
$1.000d+00$

The C matrix:

1.000d + 00 3.091d − 03 3.128d + 01 1.000d + 00 3.592d + 00 0.000d + 00

The $\hat{Q}$ matrix for the state norm:

6.637d + 00 0.0000d + 00 0.0000d + 00 0.0000d + 00 0.000d + 00 0.000d + 00
0.000d + 00 2.6554d − 07 2.6860d − 03 0.0000d + 00 3.085d − 04 0.000d + 00
0.000d + 00 2.6860d − 03 2.7174d + 01 0.0000d + 00 3.121d + 00 0.000d + 00
0.000d + 00 0.0000d + 00 0.0000d + 00 2.7174d + 01 0.000d + 00 0.000d + 00
0.000d + 00 3.0850d − 04 3.1210d + 00 0.0000d + 00 3.585d − 01 0.000d + 00
0.000d + 00 0.0000d + 00 0.0000d + 00 0.0000d + 00 0.000d + 00 0.000d + 00

The $\hat{R}$ matrix for the control norm:

5.2520d + 00

The driving noise covariance Ξ_1:

0.000d + 00 0.0000d + 00 0.0000d + 00 0.0000d + 00 0.000d + 00 0.000d + 00
0.000d + 00 0.0000d + 00 0.0000d + 00 0.0000d + 00 0.000d + 00 0.000d + 00
0.000d + 00 0.0000d + 00 0.0000d + 00 0.0000d + 00 0.000d + 00 0.000d + 00
0.000d + 00 0.0000d + 00 0.0000d + 00 0.0000d + 00 0.000d + 00 0.000d + 00
0.000d + 00 0.0000d + 00 0.0000d + 00 0.0000d + 00 1.000d − 06 0.000d + 00
0.000d + 00 0.0000d + 00 0.0000d + 00 0.0000d + 00 0.000d + 00 1.000d − 06

The measurement noise covariance matrix Ξ_2:

1.8441d − 03

The discrete-time parameters for this system sampled at 10 hertz were computed according to the equations in chapter 2:

Discrete-time transition matrix Φ (every two rows shown is actually only one row of the matrix Φ):

8.94158899875840d − 01	5.93355418621274d − 05	− 8.07897769474215d − 01
− 9.45729891346440d − 05	− 8.61683445782787d − 01	− 6.35698693844013d − 01
− 2.22006964581917d − 01	9.98658866653035d − 01	− 1.26236417743289d + 00
− 3.21783948076557d + 00	7.07041304940429d − 02	1.33003345781691d − 02
8.96632406763084d − 02	− 8.47007233273550d − 06	8.45357360609777d − 01
1.55118776302647d − 05	− 5.82375346184306d − 02	− 2.82235577782270d − 02
9.53225781519899d − 02	2.93704935200758d − 06	− 4.20099682385258d − 02
9.99996866444939d − 01	− 5.29748911536677d − 02	− 2.33843803649613d − 02
0.00000000000000d + 00	0.00000000000000d + 00	0.00000000000000d + 00
0.00000000000000d + 00	3.01194227548261d − 01	6.98805772451740d − 01
0.00000000000000d + 00	0.00000000000000d + 00	0.00000000000000d + 00
0.00000000000000d + 00	0.00000000000000d + 00	1.00000000000000d + 00

Input matrix Γ:

$-2.33843803649613d-02$
$1.22296689664596d-05$
$-8.09744421678703d-04$
$-6.19868380965549d-04$
$4.17661813028933d-02$
$9.99999956738716d-02$

State weighting matrix Q:

$6.19891054286165d+00$	$3.01509191580769d-04$	$-1.49453104023977d+00$
$1.31875722331738d+00$	$-3.31370870518672d+00$	$-1.39174639302283d+00$
$3.01509191580769d-04$	$2.47848377796048d-07$	$2.26275446201938d-03$
$2.60488981153273d-05$	$-1.83146582461534d-05$	$3.31476135012857d-05$
$-1.49453104023977d+00$	$2.26275446201938d-03$	$2.49886408802715d+01$
$-3.91315114446321d-01$	$3.00983380027126d+00$	$1.93726081288867d+00$
$1.31875722331738d+00$	$2.60488981153273d-05$	$-3.91315114446321d-01$
$2.71739589881900d+01$	$-5.29472742893936d-01$	$-1.68559768886550d-01$
$-3.31370870518672d+00$	$-1.83146582461534d-05$	$3.00983380027126d+00$
$-5.29472742893936d-01$	$2.37775667424236d+00$	$1.15440114387846d+00$
$-1.39174639302283d+00$	$3.31476135012857d-05$	$1.93726081288867d+00$
$-1.68559768886550d-01$	$1.15440114387846d+00$	$6.65122863060225d-01$

Cross-weighting matrix M:

$-3.53295371817073d-02$
$1.98100358396956d-06$
$6.49143025809568d-02$
$-3.50993906625930d-03$
$3.18190737871274d-02$
$2.03591671654107d-02$

Control weighting matrix R:

$5.25266793029727d+00$

Output matrix L:

$1.00000000000000d+00$	$3.09100000000000d-03$	$3.12800000000000d+01$
$1.00000000000000d+00$	$3.59200000000000d+00$	$0.00000000000000d+00$

State driving noise covariance matrix Θ_1:

$4.33676915174706d-08$	$-1.04991167649513d-09$	$2.00314796017624d-09$
$1.67658416889601d-09$	$-3.93401117854579d-08$	$-2.33843803650305d-08$
$-1.04991167649513d-09$	$6.13454985323572d-11$	$-6.40159235541700d-11$
$-5.69997525494007d-11$	$3.46061668931529d-10$	$1.22296689879342d-11$
$2.00314796017624d-09$	$-6.40159235541700d-11$	$9.94191359957619d-11$

$8.47200406101625d-11 \quad -1.51814214531290d-09 \quad -8.09744421672949d-10$
$1.67658416889601d-09 \quad -5.69997525494007d-11 \quad 8.47200406101625d-11$
$7.25020464385916d-11 \quad -1.20680507935355d-09 \quad -6.19868380959793d-10$
$-3.93401117854579d-08 \quad 3.46061668931529d-10 \quad -1.51814214531290d-09$
$-1.20680507935355d-09 \quad 5.93058700394746d-08 \quad 4.17661813028401d-08$
$-2.33843803650305d-08 \quad 1.22296689879342d-11 \quad -8.09744421672949d-10$
$-6.19868380959793d-10 \quad 4.17661813028401d-08 \quad 9.99999956738716d-08$

Measurement noise covariance matrix Θ_2:

$1.84412510991842d-02$

Regulator gains G (also as computed in chapter 2):

$-7.54859358895862d-01 \quad -3.38674675832647d-04 \quad 2.45537909052670d+00$
$-1.69155508507296d+00 \quad 1.04707683654752d+00 \quad 5.10491114597691d+00$

Filter gains K:

$6.30001213506085d-03$
$-2.05415833543128d-01$
$4.01197820069173d-03$
$7.47232508540808d-03$
$-2.17948949924278d-03$
$-2.17948949924279d-03$

The following tables present the data (Ψ_i, index, value) defining the 10 scaled structures analyzed in chapters 5 and 6 and the optimized structure discussed in chapter 8. Note that only the nonzero entries of the individual Ψ matrices are shown. For all the structures the output node scaling parameter ρ equals 0.02199717628337. This system also includes the A/D scale factor $k_{ad} = 0.25$ (selected to reduce overflows). This has already been taken into account in the roundoff noise results reported in chapters 5 (see section 5.2) and 8 and does not affect the coefficient wordlength results of chapter 6.

Structure (a):[a] nonzero entries in Ψ_2, Ψ_1

Matrix	Index	Value
Ψ_2	(7,1)	-2316.596730195619
Ψ_2	(7,2)	17216.30907463747
Ψ_2	(7,3)	-46538.88776849179
Ψ_2	(7,4)	60049.21454042759
Ψ_2	(7,5)	-37783.02361099942
Ψ_2	(7,6)	9373.006832979322
Ψ_2	(1,1)	1.0
Ψ_2	(2,2)	1.0
Ψ_2	(3,3)	1.0
Ψ_2	(4,4)	1.0
Ψ_2	(5,5)	1.0
Ψ_2	(6,6)	1.0
Ψ_1	(6,1)	-0.11903082227744
Ψ_1	(6,2)	1.09870649812723
Ψ_1	(6,3)	-3.98894899287426
Ψ_1	(6,4)	7.49594995996605
Ψ_1	(6,5)	-7.82430422984935
Ψ_1	(6,6)	4.33762715269116
Ψ_1	(6,8)	0.00010128626129
Ψ_1	(1,2)	1.0
Ψ_1	(2,3)	1.0
Ψ_1	(3,4)	1.0
Ψ_1	(4,5)	1.0
Ψ_1	(5,6)	1.0

a. Direct form II

Number of precedence levels: 2

Number of coefficients in scaled structure: 13

    (nonzero, nonunity entries in the modified state-space matrices)

Structure (b):[a] nonzero entries in Ψ_2, Ψ_1

Matrix	Index	Value
Ψ_2	(7,6)	0.03890104412969
Ψ_2	(7,5)	1.15283628631438
Ψ_2	(7,4)	0.13875077275467
Ψ_2	(7,3)	-0.00460563493139
Ψ_2	(7,2)	0.52228239125502
Ψ_2	(7,1)	-1.37949754700868
Ψ_2	(1,1)	1.0
Ψ_2	(2,2)	1.0
Ψ_2	(3,3)	1.0
Ψ_2	(4,4)	1.0
Ψ_2	(5,5)	1.0
Ψ_2	(6,6)	1.0
Ψ_1	(2,2)	1.46297047489118
Ψ_1	(2,1)	-0.69683507325690
Ψ_1	(6,8)	0.87673782058497
Ψ_1	(3,3)	0.99868711357757
Ψ_1	(5,8)	0.64232806309622
Ψ_1	(4,4)	0.99514095413908
Ψ_1	(4,8)	0.17364017081712
Ψ_1	(5,5)	0.58903698597208
Ψ_1	(3,8)	0.15261498391194
Ψ_1	(6,6)	0.29179162411121
Ψ_1	(2,8)	0.28980851506818
Ψ_1	(1,2)	1.0

a. Parallel direct form II, 4 first-order and 1 second-order sections
   Number of precedence levels: 2
   Number of coefficients in scaled structure: 17
      (nonzero, nonunity entries in the modified state-space matrices)

Structure (c):[a] nonzero entries in Ψ_2, Ψ_1

Matrix	Index	Value
Ψ_2	(7,6)	10.48075527883454
Ψ_2	(7,5)	-10.29571120349337
Ψ_2	(7,3)	-0.31185194361843
Ψ_2	(7,4)	0.30767918685885
Ψ_2	(7,2)	0.52228239125501
Ψ_2	(7,1)	-1.37949754700866
Ψ_2	(1,1)	1.0
Ψ_2	(2,2)	1.0
Ψ_2	(3,3)	1.0
Ψ_2	(4,4)	1.0
Ψ_2	(5,5)	1.0
Ψ_2	(6,6)	1.0
Ψ_1	(2,2)	1.46297047489119
Ψ_1	(2,1)	-0.69683507325690
Ψ_1	(4,3)	-0.29140853484973
Ψ_1	(4,4)	1.29047873768878
Ψ_1	(6,5)	-0.58617482824346
Ψ_1	(6,6)	1.58417794011116
Ψ_1	(6,8)	0.07295197592120
Ψ_1	(4,8)	0.10856479467707
Ψ_1	(2,8)	0.28980851506819
Ψ_1	(1,2)	1.0
Ψ_1	(3,4)	1.0
Ψ_1	(5,6)	1.0

a. Parallel direct form II, 3 second-order sections
Number of precedence levels: 2
Number of coefficients in scaled structure: 15
       (nonzero, nonunity entries in the modified state-space matrices)
Pole pairing (refer to table 5.1):

   z_{p1} and z_{p4}
   z_{p2} and z_{p3}
   z_{p5} and z_{p6} (these are the complex poles)

Structure (d):[a] nonzero entries in Ψ_2, Ψ_1

Matrix	Index	Value
Ψ_2	(7,6)	1.59834173340604
Ψ_2	(7,5)	-0.48730146270494
Ψ_2	(7,4)	15.71737776482720
Ψ_2	(7,3)	-15.69841756881241
Ψ_2	(7,2)	0.52228239125470
Ψ_2	(7,1)	-1.37949754700784
Ψ_2	(1,1)	1.0
Ψ_2	(2,2)	1.0
Ψ_2	(3,3)	1.0
Ψ_2	(4,4)	1.0
Ψ_2	(5,5)	1.0
Ψ_2	(6,6)	1.0
Ψ_1	(2,2)	1.46297047489118
Ψ_1	(2,1)	-0.69683507325690
Ψ_1	(4,3)	-0.99383444709201
Ψ_1	(4,4)	1.99382806771667
Ψ_1	(6,5)	-0.17187605879836
Ψ_1	(6,6)	0.88082861008328
Ψ_1	(6,8)	0.48463047627064
Ψ_1	(4,8)	0.00148815020744
Ψ_1	(2,8)	0.28980851506825
Ψ_1	(1,2)	1.0
Ψ_1	(3,4)	1.0
Ψ_1	(5,6)	1.0

a. Parallel direct form II, 3 second-order sections
   Number of precedence levels: 2
   Number of coefficients in scaled structures: 15
      (nonzero, nonunity entries in the modified state-space matrices)
   Pole pairing (Refer to table 5.1):
      z_{p1} and z_{p2}
      z_{p3} and z_{p4}
      z_{p5} and z_{p6} (these are the complex poles)

Structure (e):[a] nonzero entries in Ψ_1

Matrix	Index	Value
Ψ_1	(2,1)	-0.696835073257
Ψ_1	(2,2)	1.462970474891
Ψ_1	(2,8)	0.289808515068
Ψ_1	(7,3)	-0.089660341046
Ψ_1	(4,3)	-0.291408534850
Ψ_1	(4,4)	1.290478737689
Ψ_1	(4,8)	0.108564794677
Ψ_1	(7,4)	0.085201505052
Ψ_1	(6,5)	-0.586174828243
Ψ_1	(7,2)	-0.615413829047
Ψ_1	(6,6)	1.584177940111
Ψ_1	(6,8)	0.072951975921
Ψ_1	(7,1)	-0.363944688371
Ψ_1	(7,5)	-6.143554925433
Ψ_1	(7,6)	6.307670104940
Ψ_1	(7,8)	0.949356818741
Ψ_1	(1,2)	1.0
Ψ_1	(3,4)	1.0
Ψ_1	(5,6)	1.0

a. Parallel, one-level version of (c)
   Number of precedence levels: 1
   Number of coefficients in scaled structure: 16
        (nonzero, nonunity entries in the modified state-space matrices)
   Pole pairing: same as (c)

Structure (f):[a] nonzero entries in Ψ_1

Matrix	Index	Value
Ψ_1	(2,1)	-0.33647827003132
Ψ_1	(2,2)	0.68249200666952
Ψ_1	(2,8)	0.65051552691033
Ψ_1	(7,3)	-0.08038901173235
Ψ_1	(4,3)	-0.20036428295682
Ψ_1	(4,4)	1.19946355533120
Ψ_1	(4,8)	0.11870639047793
Ψ_1	(7,4)	0.07597348041937
Ψ_1	(6,5)	0.19085044755223
Ψ_1	(7,2)	-0.43070856151277
Ψ_1	(6,6)	0.73170685139682
Ψ_1	(6,8)	0.45237970547959
Ψ_1	(7,1)	-0.73947570074840
Ψ_1	(7,5)	-0.54742490834586
Ψ_1	(7,6)	0.94099544414975
Ψ_1	(7,8)	0.94935681874100
Ψ_1	(1,1)	0.78047846822148
Ψ_1	(1,2)	0.48789111196657
Ψ_1	(3,3)	0.09101518235780
Ψ_1	(3,4)	0.90953905526798
Ψ_1	(5,5)	0.85247108871418
Ψ_1	(5,6)	0.19692962981701
Ψ_1	(1,8)	-0.13770626781352
Ψ_1	(3,8)	-0.00287783447672
Ψ_1	(5,8)	-0.06296709412667

a. Block optimal parallel
    Number of precedence levels: 1
    Number of coefficients in scaled structure: 25
        (nonzero, nonunity entries in the modified state-space matrices)
    Pole pairing: same as (c) and (e)

Structure (g):[a] nonzero entries in Ψ_4, Ψ_3, Ψ_2, Ψ_1

Matrix	Index	Value
Ψ_4	(7,6)	-1101.542292912427
Ψ_4	(7,5)	541.2874849022007
Ψ_4	(7,7)	560.2771331108011
Ψ_4	(1,2)	1.0
Ψ_4	(2,1)	1.0
Ψ_4	(3,4)	1.0
Ψ_4	(4,3)	1.0
Ψ_4	(5,6)	1.0
Ψ_4	(6,7)	1.0
Ψ_3	(7,6)	0.88082861008329
Ψ_3	(7,5)	-0.17187605879836
Ψ_3	(7,4)	-6.00228882670692
Ψ_3	(7,3)	2.89048675941179
Ψ_3	(7,7)	3.40307257702863
Ψ_3	(1,1)	1.0
Ψ_3	(2,2)	1.0
Ψ_3	(3,7)	1.0
Ψ_3	(4,4)	1.0
Ψ_3	(5,5)	1.0
Ψ_3	(6,6)	1.0
Ψ_2	(7,3)	1.99382806771666
Ψ_2	(7,2)	-0.99383444709200
Ψ_2	(7,1)	-0.00051747678098
Ψ_2	(7,6)	0.00171808332848
Ψ_2	(1,6)	1.0
Ψ_2	(2,1)	1.0
Ψ_2	(3,2)	1.0
Ψ_2	(4,3)	1.0
Ψ_2	(5,4)	1.0
Ψ_2	(6,5)	1.0
Ψ_1	(6,2)	1.46297047489119
Ψ_1	(6,1)	-0.69683507325690
Ψ_1	(6,8)	0.28980851498215
Ψ_1	(1,2)	1.0
Ψ_1	(2,3)	1.0
Ψ_1	(3,4)	1.0
Ψ_1	(4,5)	1.0
Ψ_1	(5,6)	1.0

a. Cascade direct form II, 3 second-order sections
   Number of precedence levels: 4
   Number of coefficients in scaled structure: 15
      (nonzero, nonunity entries in the modified state-space matrices)
   Pole and zero pairing (refer to table 5.1):
      Section 1: z_{p5} and z_{p6}, z_{z1}
      Section 2: z_{p3} and z_{p4}, z_{z4} and z_{z5}
      Section 3: z_{p1} and z_{p2}, z_{z2} and z_{z3}

Structure (h):[a] nonzero entries in Ψ_4, Ψ_3, Ψ_2, Ψ_1

Matrix	Index	Value
Ψ_4	(7,6)	-35.08378898367869
Ψ_4	(7,5)	8.11873624443843
Ψ_4	(7,7)	26.98802315299709
Ψ_4	(1,2)	1.0
Ψ_4	(2,1)	1.0
Ψ_4	(3,4)	1.0
Ψ_4	(4,3)	1.0
Ψ_4	(5,6)	1.0
Ψ_4	(6,7)	1.0
Ψ_3	(7,6)	1.29047873768878
Ψ_3	(7,5)	-0.29140853484973
Ψ_3	(7,4)	-0.45738885908277
Ψ_3	(7,3)	0.22026204990314
Ψ_3	(7,7)	0.25932232325397
Ψ_3	(1,1)	1.0
Ψ_3	(2,2)	1.0
Ψ_3	(3,7)	1.0
Ψ_3	(4,4)	1.0
Ψ_3	(5,5)	1.0
Ψ_3	(6,6)	1.0
Ψ_2	(7,3)	1.46297047489118
Ψ_2	(7,2)	-0.69683507325690
Ψ_2	(7,1)	-1.79860505553554
Ψ_2	(7,6)	1.85943686039663
Ψ_2	(1,6)	1.0
Ψ_2	(2,1)	1.0
Ψ_2	(3,2)	1.0
Ψ_2	(4,3)	1.0
Ψ_2	(5,4)	1.0
Ψ_2	(6,5)	1.0
Ψ_1	(6,2)	1.58417794011116
Ψ_1	(6,1)	-0.58617482824346
Ψ_1	(6,8)	0.07295197611457
Ψ_1	(1,2)	1.0
Ψ_1	(2,3)	1.0
Ψ_1	(3,4)	1.0
Ψ_1	(4,5)	1.0
Ψ_1	(5,6)	1.0

a. Cascade direct form II, 3 second-order sections
   Number of precedence levels: 4
   Number of coefficients in scaled structure: 15
       (nonzero, nonunity entries in the modified state-space matrices)
   Pole and zero pairing (refer to table 5.1):
       Section 1: z_{p2} and z_{p3}, z_{z2}
       Section 2: z_{p5} and z_{p6}, z_{z4} and z_{z5}
       Section 3: z_{p1} and z_{p4}, z_{z1} and z_{z3}

Structure (i):[a] nonzero entries in Ψ_3, Ψ_2, Ψ_1

Matrix	Index	Value
Ψ_3	(7,3)	-320.6453445770277
Ψ_3	(7,2)	157.5620956439277
Ψ_3	(7,5)	0.88082861008329
Ψ_3	(7,4)	-0.17187605879836
Ψ_3	(7,8)	163.0897474939010
Ψ_3	(1,6)	1.0
Ψ_3	(2,1)	1.0
Ψ_3	(3,7)	1.0
Ψ_3	(4,3)	1.0
Ψ_3	(5,8)	1.0
Ψ_3	(6,5)	1.0
Ψ_2	(8,2)	-0.02595431628362
Ψ_2	(8,1)	0.01249866671420
Ψ_2	(8,4)	1.99382806771669
Ψ_2	(8,3)	-0.99383444709202
Ψ_2	(8,8)	0.01471512360540
Ψ_2	(1,2)	1.0
Ψ_2	(2,3)	1.0
Ψ_2	(3,4)	1.0
Ψ_2	(4,5)	1.0
Ψ_2	(5,6)	1.0
Ψ_2	(6,7)	1.0
Ψ_2	(7,8)	1.0
Ψ_1	(8,1)	-0.11914766720671
Ψ_1	(8,8)	0.39558416566143
Ψ_1	(8,3)	1.46297047489118
Ψ_1	(8,2)	-0.69683507325690
Ψ_1	(1,2)	1.0
Ψ_1	(2,3)	1.0
Ψ_1	(3,4)	1.0
Ψ_1	(4,5)	1.0
Ψ_1	(5,6)	1.0
Ψ_1	(6,7)	1.0
Ψ_1	(7,8)	1.0

a. Cascade direct form I, 3 second-order sections
    Number of precedence levels: 3
    Number of coefficients in scaled structure: 14
        (nonzero, nonunity entries in the modified state-space matrices)
    Pole and zero pairing: same as (g)

Structure (j):[a] nonzero entries in Ψ_3, Ψ_2, Ψ_1

Matrix	Index	Value
Ψ_3	(7,1)	0.79382319292953
Ψ_3	(7,2)	0.13324583104339
Ψ_3	(7,3)	−1.28133934418680
Ψ_3	(7,4)	1.63323383955448
Ψ_3	(7,5)	−0.22354700928633
Ψ_3	(7,6)	−1.07427890614435
Ψ_3	(1,1)	1.0
Ψ_3	(2,2)	1.0
Ψ_3	(3,3)	1.0
Ψ_3	(4,4)	1.0
Ψ_3	(5,5)	1.0
Ψ_3	(6,6)	1.0
Ψ_2	(1,1)	0.89415889987584
Ψ_2	(1,2)	0.02219872745941
Ψ_2	(1,3)	−0.40090758959435
Ψ_2	(1,4)	−0.00008683035986
Ψ_2	(1,5)	−0.17493645514225
Ψ_2	(1,6)	−0.12721034794726
Ψ_2	(2,1)	−0.00059340804849
Ψ_2	(2,2)	0.99865886665303
Ψ_2	(2,3)	−0.00167440057433
Ψ_2	(2,4)	−0.00789688283429
Ψ_2	(2,5)	0.00003836756094
Ψ_2	(2,6)	0.00000711410910
Ψ_2	(3,1)	0.18068685658836
Ψ_2	(3,2)	−0.00638575742111
Ψ_2	(3,3)	0.84535736060977
Ψ_2	(3,4)	0.00002869993983
Ψ_2	(3,5)	−0.02382581059578
Ψ_2	(3,6)	−0.01138138229375
Ψ_2	(4,1)	0.10382245521832
Ψ_2	(4,2)	0.00119679543943
Ψ_2	(4,3)	−0.02270574399998
Ψ_2	(4,4)	0.99999686644492
Ψ_2	(4,5)	−0.01171380959155
Ψ_2	(4,6)	−0.00509674027466
Ψ_2	(5,5)	0.30119422754825
Ψ_2	(5,6)	0.68880300468145
Ψ_2	(1,8)	0.25341237237753
Ψ_2	(2,8)	−0.02208549707703
Ψ_2	(3,8)	0.32520493273496
Ψ_2	(4,8)	0.32736911273784

Structure (j) (continued)

Matrix	Index	Value
Ψ_2	(5,8)	-0.43182585076519
Ψ_2	(6,8)	-0.43809680729855
Ψ_2	(1,7)	-0.02223658684666
Ψ_2	(2,7)	0.00000003108449
Ψ_2	(3,7)	-0.00155168069308
Ψ_2	(4,7)	-0.00064200329831
Ψ_2	(5,7)	0.19562955072811
Ψ_2	(6,7)	0.47519418802086
Ψ_2	(6,6)	1.0
Ψ_1	(8,2)	-0.02874919859251
Ψ_1	(8,1)	-0.02486071250568
Ψ_1	(8,3)	-0.38589414084909
Ψ_1	(8,4)	-0.02282538209825
Ψ_1	(8,5)	-0.01812935994315
Ψ_1	(8,8)	1.07470407782701
Ψ_1	(1,1)	1.0
Ψ_1	(2,2)	1.0
Ψ_1	(3,3)	1.0
Ψ_1	(4,4)	1.0
Ψ_1	(5,5)	1.0
Ψ_1	(6,6)	1.0
Ψ_1	(7,7)	1.0

a. Simple
  Number of precedence levels: 3
  Number of coefficients in scaled structure: 50
    (nonzero, nonunity entries in the modified state-space matrices)

Optimized structure considered in chapter 8 [based on structure (c)]:[a] nonzero entries in Ψ_2, Ψ_1

Matrix	Index	Value
Ψ_2	(7,6)	1.34435168286127
Ψ_2	(7,5)	−0.50777452620114
Ψ_2	(7,3)	−0.31185194361846
Ψ_2	(7,4)	0.30767918685888
Ψ_2	(7,2)	0.52228239125501
Ψ_2	(7,1)	−1.37949754700866
Ψ_2	(1,1)	1.0
Ψ_2	(2,2)	1.0
Ψ_2	(3,3)	1.0
Ψ_2	(4,4)	1.0
Ψ_2	(5,5)	1.0
Ψ_2	(6,6)	1.0
Ψ_1	(2,2)	1.46297047489118
Ψ_1	(2,1)	−0.69683507325689
Ψ_1	(4,3)	−0.29140853484973
Ψ_1	(4,4)	1.29047873768877
Ψ_1	(6,5)	0.16466105298259
Ψ_1	(6,6)	0.65028182128291
Ψ_1	(6,8)	0.56874389081747
Ψ_1	(4,8)	0.10856479467706
Ψ_1	(2,8)	0.28980851506819
Ψ_1	(5,5)	0.93389611882824
Ψ_1	(5,6)	0.12826858819766
Ψ_1	(1,2)	1.0
Ψ_1	(3,4)	1.0

a. Number of precedence levels: 2
   Number of coefficients in scaled structure: 17
   (nonzero, nonunity entries in the modified state-space matrices)

Appendix B

The Adjoint Lyapunov Operator

If we take the trace of the product of two matrices to be an inner product on the space of matrices, and π to be a matrix operator, then

$$\text{trace}(\pi(X)U) = \text{trace}(X\pi^*(U)), \qquad \text{(B.1)}$$

where π^* is the adjoint operator of π. For $\pi(X) = X - AXA'$, the operator π^* can be derived from (B.1):

$$
\begin{aligned}
\text{trace}((X - AXA')U) &= \text{trace}(XU) - \text{trace}(AXA'U) \\
&= \text{trace}(XU) - \text{trace}(XA'UA) \\
&= \text{trace}(X(U - A'UA)). \qquad \text{(B.2)}
\end{aligned}
$$

Thus $\pi^*(U) = U - A'UA$.

As used in section 5.4, the Lyapunov equation (5.38) and the trace (5.39) were replaced by the equivalent equations (5.40) and (5.41). Relating this to the derivation above gives

$$
\begin{aligned}
X &= V, \\
A &= \Psi_{11}, \\
U &= W_1, \\
\pi^*(U) &= \Pi, \\
\pi(X) &= \frac{\Delta_r^2}{12}\Lambda_1 S_1^{-2}.
\end{aligned}
\qquad \text{(B.3)}
$$

Appendix C

A Simplified Evaluation of Equation (6.23)

We shall derive here the expression used in the SWL and $MSWL$ algorithms for computing the second partial derivatives of J. Evaluating this expression will be simpler than directly computing (6.22) and (6.23). Using (6.22) and the expressions in (6.25) and (6.26), and defining the following matrices:

$$
D_1 = \left[
\begin{array}{c|c|c}
\Phi & 0_n & \Gamma\rho \\
\hline
0 & I_{n+1} & \\
\hline
L & & 0
\end{array}
\right],
\tag{C.1}
$$

$$
D_2 = \begin{bmatrix}
0_n & 0 & 0 \\
0 & 0_{n+1} & 0 \\
0 & 0 & \Theta_2
\end{bmatrix}
\tag{C.2}
$$

we can rewrite

$$
X_{ij} = \begin{bmatrix} 0 & 0 \\ 0 & \dfrac{\partial \Psi_\infty}{\partial c_j} \end{bmatrix} D_1 \frac{\partial Z}{\partial c_i} A' + \begin{bmatrix} 0 & 0 \\ 0 & \dfrac{\partial \Psi_\infty}{\partial c_i} \end{bmatrix} D_1 \frac{\partial Z}{\partial c_j} A'
$$

$$
+ \begin{bmatrix} 0 & 0 \\ 0 & \dfrac{\partial \Psi_\infty}{\partial c_i} \end{bmatrix} (D_1 Z D_1' + D_2) \begin{bmatrix} 0 & 0 \\ 0 & \dfrac{\partial \Psi_\infty'}{\partial c_j} \end{bmatrix}
$$

$$
+ \begin{bmatrix} 0 & 0 \\ 0 & \dfrac{\partial^2 \Psi_\infty}{\partial c_i \partial c_j} \end{bmatrix} (D_1 Z D_1' + D_2) \begin{bmatrix} I_n & 0 \\ 0 & \Psi_\infty' \end{bmatrix}. \tag{C.3}
$$

Thus X_{ij} will be a matrix whose lower right-hand $(n + 1) \times (n + 2)$ portion is nonzero, and the rest zero. Thus the trace expression in (6.23) can be simplified:

$$
\frac{\partial^2 J}{\partial c_i \partial c_j} = 2 \operatorname{trace} \left\{ \frac{\partial \Psi_\infty}{\partial c_j} (M1) \frac{\partial Z}{\partial c_i} (M2) + \frac{\partial \Psi_\infty}{\partial c_i} (M1) \frac{\partial Z}{\partial c_j} (M2) \right\}
$$

$$
+ 2 \operatorname{trace} \left\{ \frac{\partial \Psi_\infty}{\partial c_i} (M3) \frac{\partial \Psi_\infty}{\partial c_j} (M4) + \frac{\partial^2 \Psi_\infty}{\partial c_i \partial c_j} (M5) \right\}, \tag{C.4}
$$

where $M1$, $M2$, $M3$, $M4$, and $M5$ are precomputed matrices (computed only once for all i and j) based on the fixed matrices D_1, Z, A, D_2, $\hat{U}$, and Ψ_∞. As is shown in (C.4), a maximum of three matrix multiplications and a trace operation are required for each term in (C.4), *for each i and j*. Thus in terms of operation counts, the calculation of (6.23) would be roughly proportional to $N^2(2n + 1)^3$.

However, this expression can be further simplified to reduce the computational load. By substituting (6.27) and (6.28) for the partial derivatives $(\partial \Psi_\infty)/(\partial c_i)$, $(\partial \Psi_\infty)/(\partial c_j)$, and $(\partial^2 \Psi_\infty)/(\partial c_i \partial c_j)$, applying simple trace identities, and combining the matrices Ψ_1, Ψ_2, ..., Ψ_q with $M1$, $M2$, $M3$, $M4$, and $M5$, we can produce

$$
\frac{\partial^2 J}{\partial c_i \partial c_j} = 2 \operatorname{trace} \left\{ (M6) E_{kl} (M7) \frac{\partial Z}{\partial c_j} \right\}
$$

$$
+ 2 \operatorname{trace} \left\{ (M8) E_{rs} (M9) \frac{\partial Z}{\partial c_i} \right\}
$$

$$
+ 2 \operatorname{trace} \left\{ E_{kl} (M10) E_{rs} (M11) \right\}
$$

$$
+ 2 \operatorname{trace} \left\{ E_{ks} (M12) \right\} \qquad \text{if } l = r, \tag{C.5}
$$

where the precomputable matrices $M6$, $M7$, $M8$, $M9$, $M10$, $M11$, and $M12$ will depend on which specific precedence-level matrices contain coefficients c_i and c_j. As the number of precedence levels goes up, so does the number of such matrices—but they can still all be precomputed. Equation (C.5) can be simplified by taking advantage of the special form of E_{kl} and E_{rs} (described in section 6.3). For the first trace term of (C.5), we can write

$$(M6)E_{kl}(M7) = (V1)(V2),\tag{C.6}$$

where $V1$ is a column $(2n + 1)$-vector equal to the kth column of $M6$ and $V2$ is a row $(2n + 1)$-vector equal to the lth row of $M7$. Thus the first term of (C.5) can be written

$$2\operatorname{trace}(V1)(V2)\frac{\partial Z}{\partial c_j} = 2\operatorname{trace}(V2)\frac{\partial Z}{\partial c_j}(V1) = 2(V2)\frac{\partial Z}{\partial c_j}(V1).\tag{C.7}$$

Now only one vector matrix multiplication and one vector dot product are required to evaluate the first term of (C.5) for i and j. For all i and j, this term can thus be evaluated with a number of computations roughly proportional to $N^2(2n + 1)^2$.

The second term of (C.5) can be simplified in exactly the same manner as the first. The third term, since there is no dependence on c_i or c_j other than in E_{kl} and E_{rs}, can be reduced to

$$2\operatorname{trace}\{E_{kl}(M10)E_{rs}(M11)\} = 2(M10(l,r)M11(s,k)).\tag{C.8}$$

This involves even less computation then the first two terms. Finally, the fourth term reduces to the simplest form of all:

$$2\operatorname{trace}\{E_{ks}(M12)\} = 2(M12(s,k)).\tag{C.9}$$

Thus overall, the number of operations involved in computing (6.23), assuming that we already have the first partial derivatives of Z, has been reduced from being roughly proportional to $N^2(2n + 1)^3$ to being roughly proportional to $N^2(2n + 1)^2$, where N is the number of rounded coefficients in the structure and n is the order of the plant.

References

1. H. Kwakernaak and R. Sivan, *Linear Optimal Control Systems*, J. Wiley & Sons, New York, 1972.

2. J. C. Willems and S. K. Mitter, "Controllability, Observability, Pole Allocation, and State Reconstruction," *IEEE Trans. Aut. Control*, Vol. AC-16, No. 6, December 1971, pp. 582–595.

3. M. Athans, "The Discrete Time Linear-Quadratic-Gaussian Stochastic Control Problem," *Annals of Economic and Social Measurement*, Vol. 1, No. 4, 1972, pp. 449–491.

4. B. C. Kuo, *Analysis and Synthesis of Sampled-Data Control Systems*, Prentice-Hall, Englewood Cliffs, New Jersey, 1963.

5. D. M. Auslander, Y. Takahashi, and M. Tomizuka, "Direct Digital Process Control: Practice and Algorithms for Microprocessor Application," *Proc. IEEE*, Vol. 66, No. 2, February 1978, pp. 199–208.

6. T. F. Tao, D. Bar Yehoshua, and R. Martinez, "Applications of Microprocessors in Control Problems," *Proc. 1977 Joint Automatic Control Conf.*, 1977, pp. 8–13.

7. J. B. Knowles and R. Edwards, "Effect of a Finite-Word-Length Computer in a Sampled-Data Feedback System," *Proc. IEE*, Vol. 112, No. 6, June 1965, pp. 1197–1207.

8. E. E. Curry, "The Analysis of Round-Off and Truncation Errors in a Hybrid Control System," *IEEE Trans. Aut. Control*, Vol. AC-13, October 1967, pp. 601–604.

9. J. E. Bertram, "The Effect of Quantization in Sampled-Feedback Systems," *Trans. Amer. Inst. Elec. Engrs.*, Vol. 77, Pt. 2, September 1958, pp. 177–182.

10. J. B. Slaughter, "Quantization Errors in Digital Control Systems," *IEEE Trans. Aut. Control*, Vol. AC-9, No. 1, January 1964, pp. 70–74.

11. G. W. Johnson, "Upper Bound on Dynamic Quantization Error in Digital Control Systems via the Direct Method of Liapunov," *IEEE Trans. Aut. Control*, Vol. AC-10, No. 4, October 1965, pp. 439–448.

12. G. N. T. Lack and G. W. Johnson, "Comments on "Upper Bound On Dynamic Quantization Error in Digital Control Systems Via the Direct Method of Liapunov"," *IEEE Trans. Aut. Control*, Vol. AC-11, April 1966, pp. 331–334.

13. A.B. Sripad, "Models for Finite Precision Arithmetic, With Application to the Digital Implementation of Kalman Filters," Sc. D. Dissertation, Washington Univ., Sever Institute, January 1978.

14. R. E. Rink and H. Y. Chong, "Performance of State Regulator Systems with Floating-Point Computation," *IEEE Trans. Aut. Control*, Vol. AC-24, No. 3, June 1979, pp. 411–421.

15. F. A. Farrar, "Microprocessor Implementation of Advanced Control Modes," *Summer Computer Simulation Conference Proceedings*, Chicago, Illinois, July 1977, pp. 339–342.

16. A. S. Willsky, "Digital Signal Processing and Control and Estimation Theory— Points of Tangency, Areas of Intersection, and Parallel Directions," MIT ESL Rept. ESL-R-712, Cambridge, Mass., January 1977 (also published by MIT Press, 1979).

17. D. S. K. Chan, "Theory and Implementation of Multidimensional Discrete Systems for Signal Processing," Ph.D. Dissertation, MIT, Department of Electrical Engineering and Computer Science, May 1978.

18. C. T. Mullis and R. A. Roberts, "Synthesis of Minimum Roundoff Noise Fixed-Point Digital Filters," *IEEE Trans. Circuits & Systems*, Vol. CAS-23, No. 9, September 1976, pp. 551–562.

19. J. B. Knowles and E. M. Olcayto, "Coefficient Accuracy and Digital Filter Response," *IEEE Trans. Circuits & Systems*, Vol. CAS-15, March 1968, pp. 31–41.

20. E. Avenhaus, "On the Design of Digital Filters with Coefficients of Limited Word Length," *IEEE Trans. Audio & Electroacoustics*, Vol. AU-20, August 1972, pp. 206–212.

21. R. E. Crochiere, "A New Statistical Approach to the Coefficient Word Length Problem for Digital Filters," *IEEE Trans. Circuits & Systems*, Vol. CAS-22, No. 3, March 1975, pp. 190–196.

22. J. F. Kaiser, "On the Limit Cycle Problem," *Proc. IEEE Inter. Conf. Acous. Speech & Signal Processing*, 1976, pp. 642–644.

23. M. Athans, guest ed., *IEEE Transactions Aut. Control, Special Issue on Linear-Quadratic-Gaussian Problem*, Vol. AC-16, No. 6, December 1971.

24. M. G. Safonov and M. Athans, "Gain and Phase Margin for Multiloop LQG

Regulators," *Proc. IEEE Trans. Aut. Control*, Vol. AC-22, No. 2, April 1977, pp. 173–179.

25. C. A. Harvey and G. Stein, "Quadratic Weights for Asymptotic Regulator Properties," *Proc. 1977 IEEE Conf. Decision & Control*, Vol. 1, 1977, pp. 1220–1228.

26. G. Stein, "Generalized Quadratic Weights for Asymptotic Regulator Properties," *IEEE Transactions Aut. Control*, Vol. AC-24, No. 4, August 1979, pp. 559–566.

27. G. K. Roberts, "Consideration of Computer Limitations in Implementing On-Line Controls," MIT ESL Rept. ESL-R-665, Cambridge, Mass. June 1976.

28. A. Gelb, ed., *Applied Optimal Estimation*, MIT Press, Cambridge, Mass., 1974.

29. A. V. Oppenheim and R. W. Schafer, *Digital Signal Processing*, Prentice Hall, Inc., Englewood Cliffs, New Jersey, 1975.

30. L. R. Rabiner and B. Gold, *Theory and Application of Digital Signal Processing*, Prentice Hall, Inc., Englewood Cliffs, New Jersey, 1975.

31. S. A. Tretter, *Introduction to Discrete-Time Signal Processing*, J. Wiley & Sons, New York, New York, 1976.

32. A. P. Sage, *Optimal Systems Control*, Prentice-Hall, Englewood Cliffs, New Jersey, 1968.

33. R. E. Crochiere and A. V. Oppenheim, "Analysis of Linear Digital Networks," *Proc. IEEE*, Vol. 63, No. 4, April 1975, pp. 581–595.

34. R. E. Crochiere, "Digital Network Theory and Its Application to the Analysis and Design of Digital Filters," Ph.D. Dissertation, MIT, Department of Electrical Engineering, April 1974.

35. J. F. Kaiser, "Some Practical Considerations in the Realization of Linear Digital Filters," *Proc. Third Annual Allerton Conf. Circuit & System Theory*, October 1965, Monticello, Illinois, pp. 621–633.

36. E. Avenhaus, "A Proposal to Find Suitable Canonic Structures for the Implementation of Digital Filters with Small Coefficient Wordlength," *Nachr. Tech. Zeit.*, Vol. 25, No. 8, August 1972, pp. 377–382.

37. E. Lueder and K. Haug, "Calculation of All Equivalent and Canonic 2nd Order Digital Filter Structures," *Proc. 1978 IEEE Inter. Conf. Acous. Speech & Signal Processing*, Tulsa, Oklahoma, April 1978, pp. 51–54.

38 R. W. Brockett, *Finite Dimensional Linear Systems*, J. Wiley and Sons, Inc., New York, New York, 1970.

39. C. T. Mullis and R. A. Roberts, "Roundoff Noise in Digital Filters—Frequency Transformations and Invariants," *IEEE Trans. Acous. Speech & Signal Processing*, Vol. ASSP-24, No. 6, December 1976, pp. 538–550.

40. C. T. Mullis and R. A. Roberts, "Filter Structures Which Minimize Roundoff Noise in Fixed-Point Digital Filters," *Proc. IEEE Inter. Conf. Acous. Speech & Signal Processing*, April 1976, Philadelphia, Penn., pp. 505–508.

41. S. Y. Hwang, "Minimum Uncorrelated Unit Noise in State-Space Digital Filters," *IEEE Trans. Acous. Speech & Signal Processing*, Vol. ASSP-25, No. 4, August 1977, pp. 273–281.

42. L. B. Jackson, A. G. Lindgren, and Y. Kim, "Synthesis of State-Space Digital Filters with Low Roundoff Noise and Coefficient Sensitivity," *Proc. IEEE Inter. Symp. Circuits & Systems*, April 1977, Phoenix, Arizona, pp. 41–44. (also *IEEE Trans. Circuits & Systems*, Vol. CAS-26, No. 3, March 1979, pp. 149–153.)

43. A. H. Gray, Jr., and J. D. Markel, "Digital Ladder and Lattice Filter Synthesis," *IEEE Trans. Audio & Electroacoustics*, Vol. AU-21, No. 6, December 1974, pp. 491–500.

44. S. K. Mitra and R. J. Sherwood, "Canonic Realizations of Digital Filters Using the Continued Fraction Expansion," *IEEE Trans. Audio & Electroacoustics*, Vol. AU-20, No. 3, August 1972, pp. 185–194.

45. A. Fettweis, "Digital Filter Structures Related to Classical Filter Networks," *Arch. Elek. Übert.*, Vol. 25, No. 2, February 1971, pp. 79–89.

46. R. E. Crochiere, "Digital Ladder Structures and Coefficient Sensitivity," *IEEE Trans. Audio & Electroacoustics*, Vol. AU-20, No. 4, October 1972, pp. 240–246.

47. K. Renner and S. C. Gupta, "On the Design of Wave Digital Filters With Low Sensitivity Properties," *IEEE Trans. Circuit Theory*, Vol. CT-20, No. 5, September 1973, pp. 555–567.

48. A. Fettweis, "Pseudopassivity, Sensitivity, and Stability of Wave Digital Filters," *IEEE Trans. Circuit Theory*, Vol. CT-19, No. 6, November 1972, pp. 668–673.

49. A. Fettweis and K. Meerkötter, "Suppression of Parasitic Oscillations in Wave Digital Filters," *IEEE Trans. Circuits & Systems*, Vol. CAS-22, March 1975, pp. 239–246, and June 1975, p. 575.

50. A. Fettweis, "On Sensitivity and Roundoff Noise in Wave Digital Filters," *IEEE Trans. Acous. Speech & Signal Processing*, Vol. ASSP-22, No. 5, October 1974, pp. 383–384.

51. A. Sedlmeyer and A. Fettweis, "Digital Filters With True Ladder Configuration," *Inter. Jour. of Circuit Theory and Applic.*, Vol. 1, No. 1, March 1973, pp. 5–10.

52. A. Fettweis, "Wave Digital Filters with Reduced Number of Delays," *Inter. Jour. of Circuit Theory and Applic.*, Vol. 2, No. 4, 1974, pp. 319–320.

53. K. Meerkötter and W. Wegener, "A New Second-Order Digital Filter Without Parasitic Oscillations," *Arch. Elek. Übert.*, Vol. 29, No. 7/8, July/August 1975, pp. 312–314.

54. J. Allen and R. G. Gallagher, *Computation Structures*, MIT Course Notes for 6.032, 1977.

55. S. K. Tewksbury, R. B. Kieburtz, J. S. Thompson, and S. P. Verma, "Tutorials on Signal Processing for Communications: Part II-Digital Signal Processing Architecture," *IEEE Communications Society Magazine*, January 1978, pp. 23–27.

56. J. Allen, "Computer Architecture for Signal Processors," *Proc. IEEE*, Vol. 63, No. 4, April 1975, pp. 624–633.

57. L. B. Jackson, "On the Interaction of Roundoff Noise and Dynamic Range in Digital Filters," *Bell Syst. Tech. Journal*, Vol. 49, No. 2, February 1970, pp. 159–184.

58. S. Y. Hwang, "Dynamic Range Constraints in State-Space Digital Filtering," *IEEE Trans. Acous. Speech & Signal Processing*, Vol. ASSP-23, No. 6, December 1975, pp. 591–593.

59. A. V. Oppenheim and C. J. Weinstein, "Effects of Finite Register Length in Digital Filtering and the Fast Fourier Transform," *Proc. IEEE*, Vol. 60, August 1972, pp. 957–976.

60. A. B. Sripad and D. L. Snyder, "Necessary and Sufficient Conditions for Quantization Errors to be Uniform & White," *IEEE Trans. Acous. Speech & Signal Processing*, Vol. ASSP-25, No. 5, October 1977, pp. 442–448.

61. T. A. C. M. Claasen, W. F. G. Mecklenbräuker, and J. B. H. Peek, "Quantization Noise Analysis for Fixed-Point Digital Filters Using Magnitude Truncation," *IEEE Trans. Circuit & Systems*, Vol. CAS-22, No. 11, Nov. 1975, pp. 887–895.

62. T. A. C. M. Claasen, W. F. G. Mecklenbräuker, and J. B. H. Peek, "Effects of Quantization and Overflow in Recursive Digital Filters," *IEEE Trans. Acous. Speech & Signal Processing*, Vol. ASSP-24, No. 6, December 1976, pp. 517–529.

63. L. B. Jackson, "Roundoff Noise Analysis for Fixed-Point Digital Filters Realized in Cascade or Parallel Form," *IEEE Trans. Audio & Electroacoustics*, Vol. AU-18, June 1970, pp. 107–122.

64. S. K. Mitra, K. Hirano, and H. Sakaguchi, "A Simple Method of Computing the Input Quantization and Multiplication Roundoff Errors in a Digital Filter," *IEEE Trans. Acous. Speech & Signal Processing*, Vol. ASSP-22, No. 5, October 1974, pp. 326–329.

65. S. Y. Hwang, "Roundoff Noise in State-Space Digital Filtering: A General Analysis," *IEEE Trans. Acous. Speech & Signal Processing*, Vol. ASSP-24, No. 3, June 1976, pp. 256–262.

66. A. E. Bryson, Jr., guest ed. Mini-Issue on the F-8 DFBW, *IEEE Trans. Aut. Control*, Vol. AC-22, No. 5, October 1977, pp. 752–806.

67. G. Dehner, "A Contribution to the Optimization of Roundoff-Noise in Recursive Digital Filters," *Arch. Elek. Übert.*, Vol. 29, No. 12, December 1975, pp. 505–510.

68. S. Y. Hwang, "On Optimization of Cascade Fixed-Point Digital Filters," *IEEE Trans. Circuit & Systems*, Vol. CAS-21, No. 1, January 1974, pp. 163–165.

69. K. Steiglitz, "Designing Short-word Recursive Digital Filters," *Proc. 9th Allerton Conf.*, October 1971, pp. 778–788.

70. N. I. Smith, "A Random-Search Method for Designing Finite-Wordlength Recursive Digital Filters," *IEEE Trans. Acous. Speech & Signal Processing*, Vol. ASSP-27, No. 1, February 1979, pp. 40–46.

71. C. M. Rader and B. Gold, "Effects of Parameter Quantization on the Poles of a Digital Filter," *Proc. IEEE*, Vol. 55, May 1967, pp. 688–689.

72. A. I. Abu-El-Haija, K. Shenoi, and A. M. Peterson, "A Structure Suitable for Implementing Digital Filters with Poles Near +1," *National Telecommunication Conf.*, 1977, pp. 29:5-1–29:5-7.

73. R. C. Agarwal and C. S. Burrus, "New Recursive Digital Filter Structures Having Low Sensitivity and Roundoff Noise," *IEEE Trans. Circuit & Systems*, Vol. CAS-22, No. 12, December 1975, pp. 921–927.

74. D. S. K. Chan and L. R. Rabiner, "Analysis of Quantization Errors in the Direct Form for Finite Impulse Response Digital Filters," *IEEE Trans. Audio & Electroacoustics*, Vol. AU-21, August 1973, pp. 354–366.

75. W. S. Levine, T. L. Johnson, and M. Athans, "Optimal Limited State Variable Feedback Controllers for Linear Systems," *IEEE Trans. Aut. Control*, Vol. AC-16, No. 6, December 1971, pp. 785–793.

76. D. P. Looze, P. K. Houpt, N. R. Sandell, Jr., and M. Athans, "On Decentralized Estimation and Control with Application to Freeway Ramp Metering," *IEEE Trans. Aut. Control*, Vol. AC-23, No. 2, April 1978, pp. 268–275.

77. A. Y. Barraud, "A Numerical Algorithm to Solve $A^T X A - X = Q$," *IEEE Trans. Aut. Control*, Vol. AC-22, No. 5, October 1977, pp. 883–885.

78. A. Fettweis, "Roundoff Noise and Attenuation Sensitivity in Digital Filters with Fixed-Point Arithmetic," *IEEE Trans. Circuit Theory*, Vol. CT-20, No. 2, March 1973, pp. 174–175.

79. A. Fettweis, "On the Connection Between Multiplier Wordlength Limitations and Roundoff Noise in Digital Filters," *IEEE Trans. Circuit Theory*, Vol. CT-19, No. 5, September 1972, pp. 486–491.

80. R. B. Kieburtz, "An Experimental Study of Roundoff Effects in a 10th-Order Recursive Digital Filter," *IEEE Trans. Communications*, Vol. COM-21, No. 6, June 1973, pp. 757–763.

81. T. A. C. M. Claasen, W. F. G. Mecklenbräuker, and J. B. H. Peek, "Frequency-Domain Criteria for the Absence of Zero-Input Limit Cycles in Nonlinear Discrete-Time Systems, With Applications to Digital Filters," *IEEE Trans. Circuit & Systems*, Vol. CAS-22, No. 3, March 1975, pp. 232–239.

82. D. D. Šiljak, "Algebraic Criteria for Positive Realness Relative to the Unit Circle," *J. Franklin Inst.*, Vol. 296, No. 2, August 1973, pp. 115–122.

83. W. L. Mills, C. T. Mullis, and R. A. Roberts, "Digital Filters Without Overflow Oscillations," *IEEE Trans. Acous. Speech & Signal Processing*, Vol. ASSP-26, No. 4, August 1978, pp. 334–338.

84. C. W. Barnes and A. T. Fam, "Minimum Norm Recursive Digital Filters that Are Free of Overflow Limit Cycles," *IEEE Trans. Circuits & Systems*, Vol. CAS-24, No. 10, October 1977, pp. 569–574.

85. C. W. Barnes, "Roundoff Noise and Overflow in Normal Digital Filters," *IEEE Trans. Circuit & Systems*, Vol. CAS-26, No. 3, March 1979, pp. 154–159.

86. A. T. Fam and C. W. Barnes, "Nonminimal Realizations of Fixed-Point Digital Filters That Are Free of All Finite Word-Length Limit Cycles," *IEEE Trans. Acous. Speech & Signal Processing*, Vol. ASSP-27, No. 2, April 1979, pp. 149–153.

87. L. B. Jackson, "Limit Cycles in State-Space Structures for Digital Filters," *IEEE Trans. Circuit & Systems*, Vol. CAS-26, No. 1, January 1979, pp. 67–68.

88. S. R. Parker and S. F. Hess, "Limit Cycle Oscillations in Digital Filters," *IEEE Trans. Circuit Theory*, Vol. CT-18, No. 6, November 1971, pp. 687–697.

89. L. B. Jackson, "An Analysis of Limit Cycles Due to Multiplicative Rounding in Recursive Digital Filters," *Proc. 7th Allerton Conf. Circuit & System Theory*, Monticello, Illinois, October 1969, pp. 69–78.

90. I. W. Sandberg and J. F. Kaiser, "A Bound on Limit Cycles in Fixed-Point Implementations of Digital Filters," *IEEE Trans. Audio & Electroacoustics*, Vol. AU-20, No. 2, June 1972, pp. 110–112.

91. R. B. Kieburtz, V. B. Lawrence, and K. V. Mina, "Control of Limit Cycles in Recursive Digital Filters By Randomized Quantization," *IEEE Trans. Circuit & Systems*, Vol. CAS-24, No. 6, June 1977, pp. 291–299.

92. G. Zames and N. A. Shneydor, "Dither in Nonlinear Systems," *IEEE Trans. Aut. Control*, Vol. AC-21, No. 5, October 1976, pp. 660–667.

93. G. Zames and N. A. Shneydor, "Structural Stabilization and Quenching by Dither in Nonlinear Systems," *IEEE Trans. Aut. Control*, Vol. AC-22, No. 3, June 1977, pp. 352–361.

94. V. B. Lawrence and K. V. Mina, "Control of Limit Cycle Oscillations in Second-Order Recursive Digital Filters Using Constrained Random Quantization," *IEEE Trans. Acous. Speech & Signal Processing*, Vol. ASSP–26, No. 2, April 1978, pp. 127–134.

95. M. Büttner, "Elimination of Limit Cycles in Digital Filters with Very Low Increase in Quantization Noise," *IEEE Trans. Circuits & Systems*, Vol. CAS-24, No. 6, June 1977, pp. 300–304.

96. A. N. Willson, Jr., "Limit Cycles Due to Adder Overflow in Digital Filters," *IEEE Trans. Circuit Theory*, Vol. CT-19, No. 4, July 1972, pp. 342–346.

97. P. M. Ebert, J. E. Mazo, and M. G. Taylor, "Overflow Oscillations in Digital Filters," *Bell Syst. Tech. Journal*, Vol. 48, No. 9, Nov. 1969, pp. 2999–3020.

98. T. A. C. M. Claasen, W. F. G. Mecklenbräuker, and J. B. H. Peek, "On the Stability of the Forced Response of Digital Filters with Overflow Nonlinearities," *IEEE Trans. Circuits & Systems*, Vol. 22, No. 8, August 1975, pp. 692–696.

99. T. A. C. M. Claasen and L. Kristiansson, "Necessary and Sufficient Conditions for the Absence of Overflow Oscillations in 2nd-Order Recursive Digital Filters,"

IEEE Trans. Acous. Speech & Signal Processing, Vol. ASSP-23, No. 6, December 1975, pp. 509–515.

100. A. N. Willson, Jr., "Some Effects of Quantization and Adder Overflow on the Forced Response of Digital Filters," *Bell Syst. Tech. Journal*, Vol. 51, No. 4, April 1972, pp. 863–887.

101. A. Fettweis and K. Meerkötter, "On Parasitic Oscillation in Digital Filters Under Looped Conditions," *IEEE Trans. Circuit and Systems*, Vol. CAS-24, No. 9, September 1977, pp. 475–481.

102. D. S. K. Chan, "Constrained Minimization of Roundoff Noise in Fixed-Point Digital Filters," *Proc. 1979 IEEE Inter. Conf. Acous. Speech & Signal Processing*, Washington, D. C., April 1979, pp. 335–339.

103. S. M. Selby, ed., *CRC Standard Math Tables*, 19th edition, The Chemical Rubber Co., Cleveland, Ohio, 1971.

104. S. R. Parker and P. E. Girard, "Correlated Noise Due to Roundoff in Fixed Point Digital Filters," *IEEE Trans. Circuit and Systems*, Vol. CAS-23, No. 4, April 1976, pp. 204–211.

105. A. D. Edgar and S. C. Lee, "FOCUS Microcomputer Number System," *Communications of the ACM*, Vol. 22, No. 3, March 1979, pp. 166–177.

106. T. L. Johnson, "Finite-State Compensators for Physical Systems," MIT ESL Technical Memorandum ESL-TM-658, Cambridge, Mass., April 1976.

Index